Adaptations of Desert Organisms

Edited by J.L. Cloudsley-Thompson

Springer
Berlin
Heidelberg
New York
Barcelona
Budapest
Hong Kong
London
Milan
Paris
Singapore
Tokyo

George P. Stamou

Arthropods of Mediterranean-Type Ecosystems

With 72 Figures and 16 Tables

Springer

Prof. Dr. George P. Stamou
Department of Ecology
School of Biology
Faculty of Sciences
Aristotle University
U.P. Box 119
540 06 Thessaloniki
Greece

Front cover illustration:
Ventral view of *Glomeris balcanica* (millipede). Photograph taken by Dr. G.D. Iatrou.

ISBN-13: 978-3-642-79754-5   e-ISBN-13: 978-3-642-79752-1
DOI: 10.1007/978-3-642-79752-1

Library of Congress Cataloging–in–Publication Data

Stamou, George P., 1947- Arthropods of Meditterranean-type ecosystems / George P. Stamou. p. cm. -- (Adaptations of desert organisms, ISSN 1430432) Includes bibliographical references (p.   ) and index.   ISBN-13: 978-3-642-79754-5   1. Arthropoda. 2. Mediterranean-type ecosystems. I. Title. II. Series. QL434.S87 1998 595--dc21 98-3139

Cover design: *design & production* GmbH, Heidelberg
Camera ready by: Emanuel Rachl

SPIN 10476059   31/3136 - 5 4 3 2 1 0 - Printed on acid-free paper

To J.P. Cancela Da Fonseca

Preface

It is generally realised that knowledge of the adaptations of Mediterranean arthropods is limited and almost entirely restricted to qualitative aspects of the subject. This probably stems from the fact that Mediterranean arthropods were originally considered to be part of an intermediate ecosystem situated between arid and temperate ecosystems. Within these confines, the comparison of qualitative adaptations with those exhibited by arid, temperate and arctic/antarctic arthropods allowed for easy assignment of adaptive values to different characteristics. The focal point of this book, however, is quite different. The extremity of the Mediterranean environment is the result of neither cold nor drought but of a varied landscape, where temporal fluctuations in climatic variables as well as the impacts of human activities are nevertheless largely predictable. Accordingly, Mediterranean arthropods exhibit specific adaptations in response to spatial gradients as well as diurnal and interannual, but mainly seasonal fluctuations in temperature and humidity.

The adaptations of Mediterranean arthropods merit discussion with respect to the physiological, behavioural and historical constraints driving habitat selection, compartmentalisation of activity, switching activity, seasonal life cycles etc. In other words, the adaptations of Mediterranean arthropods are discussed within a specific context. In this context, traditional comparisons appear less fruitful. They may lead to misleading conclusions and have been limited in this book unless they contribute significantly to my arguments.

The second cornerstone of the above argument is that, in order to organise ideas within a strategic context, it is necessary to quantify the data. Thus, the discussion of different topics begins with a presentation of the information available, which is mainly qualitative followed by commendation of the quantitative aspects of the subject matter.

My adventure with the study of Mediterranean-type ecosystems started intensively in a temperate forest environment at Fontainebleau, France, almost 20 years ago. Dr. J.P. Cancela Da Fonseca introduced me to ecological thinking. This book is devoted to him. Many thanks are also due to Professor N.S. Margaris who first brought me into contact with Mediterranean-type ecosystems. I feel profoundly indebted to Assistant Professors S.P. Sgardelis and J.D. Pantis. Ambitious young ecologists at the time, we passionately debated the nature of "Mediterraneity". It was indeed a fruitful period. I am also indebted to my collaborators and colleagues G.D. Iatrou, M.D. Asikidis, M.D. Argyropoulou, K.J. Korfiatis and E.M. Papatheodorou, as well as to my good friend Angela, the secretary of our department. Their contribution to the elaboration of concrete ideas concerning Mediterranean-type ecosystems is invaluable. Finally, many thanks to Dr. J.L. Cloudsley-Thompson who gave me the opportunity to write this book, commented on earlier drafts and checked the text linguistically.

Paleokastro, March 1998   G. P. Stamou

Contents

1	**Introduction**	1
1.1	The Mediterranean Climate	1
1.2	Mediterranean-Type Ecosystems	2
1.2.1	The Physical Environment	2
1.2.2	Vegetation	6
1.2.3	Geographical Distribution	7
1.2.3.1	Maquis Formations	7
1.2.3.2	Phrygnic Formations	9
1.2.4	The Resilience of Mediterranean-Type Ecosystems	10
1.2.4.1	Attributes of Resilience	13
1.2.5	Litter Production and Decomposition	15
2	**Water Balance in Mediterranean Arthropods**	17
2.1	Water Relations	17
2.2	Components of Water Balance	18
2.3	The Dynamics of Water Relations	19
2.4	Adaptation to Moisture Variations	22
2.5	Responses of Arthropods to Environmental Extremes	24
2.5.1	Behavioural Responses	24
2.5.2	Cryptobiosis	25
2.5.2.1	Anhydrobiosis	25
2.5.2.2	Ecomorphosis	27
3	**Respiratory Metabolism**	29
3.1	Respiration and Weight	30
3.2	Respiratory Response to Varying Temperature	32
3.3	Acclimation to Constant Temperature	35

4   Activity Patterns 41

4.1   Daily Patterns 41
4.2   Seasonal Patterns 42
4.3   Feeding Activity 44
4.4   Oviposition Patterns 46
4.5   The Joint Effect of Temperature and Humidity
          on Feeding Activity and Demography 49
4.5.1   Short-Term Effect of Varying Temperature
          and Humidity 49
4.5.2   Energetics 51
4.5.3   Long-Term Effect of Temperature 53
4.5.4   The Effect of Thermal Past
          on Demographic Parameters 55

5   Life Cycle Tactics and Development 59

5.1   Life History Characteristics 59
5.1.1   Life Span 61
5.1.2   Age of Maturity 62
5.1.3   Reproductive Strategies 62
5.1.3.1   Thelytoky and Sexual Reproduction 63
5.1.3.2   Semelparity and Iteroparity 65
5.1.3.3   Parental Care 67
5.1.3.4   Reproductive Effort 68
5.1.3.5   The Effect of Temperature and Humidity
          on Egg-Laying Patterns 68
5.2   Synchronisation Tactics 69
5.3   Life Cycle Development 71
5.3.1   Life Cycle Development
          of Short-Lived Arthropods 71
5.3.2   Life Cycle Development
          of Long-Lived Arthropods 73

6   Phenological Patterns 77

6.1   Modelling Phenological Patterns 77
6.2   Numerical Responses of Microarthropods 79
6.3   Numerical Responses of Macroarthropods 84
6.4   Numerical Responses to Environmental Disasters .. 86
6.4.1   Population Dynamics 86
6.4.2   Modelling Population Dynamics 87

7	**Community Structure**	91
7.1	The Composition of Arthropod Communities	91
7.2	Seasonal Variations in Numbers	93
7.3	Spatial Patterns	96
7.3.1	Vertical Movement	96
7.3.2	Horizontal Distribution	97
7.4	Community Structures Induced by Human Practices	99
7.4.1	Fire-Induced Structures	100
7.4.1.1	The Effect of Fire on Microarthropods	100
7.4.1.2	The Effect of Fire on Macroarthropods	103
7.4.1.2.1	Numerical Response	103
7.4.1.2.2	Community Composition	105
7.4.2	Grazing Induced Structures	106
7.5	The Biotic Correlates of Habitat Selection	110
8	**Synthesis**	117
References		125
Species Index		137
Subject Index		139

CHAPTER 1

Introduction

1.1
The Mediterranean Climate

In addition to the regions surrounding Mediterranean Sea, a Mediterranean climate is also typical of the western coasts of the five continents, i.e. most of California, central Chile, parts of south-western South Africa and parts of south-western Australia (Aschman 1973; Daget 1984). According to Aschman, a Mediterranean climate is characterised by a hot summer with a high rate of evapotranspiration and a mild winter with rather high precipitation. Aschman defined the Mediterranean climate quantitatively as follows: annual precipitation in the range of 275 to 900 mm with over 65% occurring during the 6 colder months. During this period the monthly average temperature is below 15 °C, while the temperature falls below freezing for less than 3% of the year.

Regions characterised by a Mediterranean climate occupy less than 5% of the earth's surface between 25° and 45° latitude north and south (Tsiourlis 1990). Traditionally, most ecological studies in the Mediterranean regions focus on the structure and function of vegetation. As a consequence, the definition of Mediterranean bioclimatic zones is based primarily upon variables associated with plant growth, such as fluctuating temperature and, most notably, the amount and seasonal distribution of rainfall. Seasonality in precipitation is higly characteristic of the Mediterranean climate, and Lamotte and Bladin (1989) contrasted the regularly occurring summer drought in Mediterranean areas with the irregular distribution of rainfall in temperate regions, as well as with rainfall occurring during a limited time period in tropical regions.

Sauvage (1961) and, more recently, Nahal (1981) used Ebergers' index of drought to subdivide Mediterranean climate into subclasses as follows:

$$Q = \frac{2P}{(M+m)(M-m)},$$

where P is annual precipitation, M the mean maximum temperature of the hottest month, and m the mean minimum temperature of the coldest month.

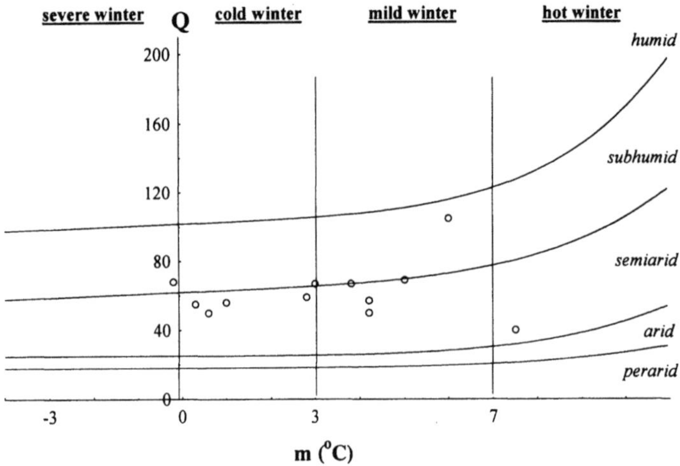

Fig 1.1. Bioclimatic classification of 12 Greek Mediterranean areas. *m* Mean minimum temperature of the coldest month, Q Eberger's pluviometric quotient (Data from Mardiris 1992)

Nahal provided a graph in which Q was plotted against m. The reason for using m is that minimum temperature in winter defines a biological threshold for plant growth. Five subclasses of the Mediterranean climate were distinguished: humid, subhumid, semi-arid, arid and perarid. In Fig. 1.1 the bioclimatic classification of 12 Greek areas of maquis is given. As can be seen, most areas fall in the semi-arid bioclimatic class.

Beyond more or less conventional definitions, exact definition of the "Mediterranean" concept remains controversial (Lamotte and Bladin 1989). For some authors the term "Mediterranean" is associated with a clear-cut drought period followed by a humid one (Di Castri 1973; Quezel and Barbero 1982), while other authors assign this term to situations combining summer drought and winter cold stresses (Mitrakos 1980).

1.2
Mediterranean-Type Ecosystems

1.2.1
The Physical Environment

Most worldwide areas with a Mediterranean climate fall into the semi-arid bioclimatic class. Considering cost-benefit ratios, in which cost refers to leaf production and benefit to carbon gain, Mooney (1977) suggested that evergreen-sclerophyllous formations can be found in regions with shorter drought periods. These formations are known as maquis (Mediterranean countries),

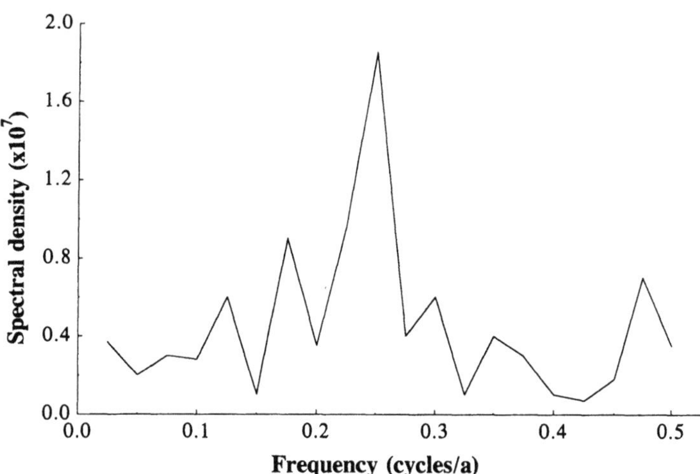

Fig. 1.2. Periodogram analysis of Emberger's pluviometric index estimated for the period 1946-1984 (Data from Iatrou 1989)

matoral (Chile), chaparral (California), fynbos (South Africa), and mallee (Australia). The dry end point of the Mediterranean humidity gradient is occupied by formations dominated by seasonally dimorphic, drought deciduous shrubs. These formations are known in Greece as phrygana, in Israel as batha, in Spain as tomillares, in California as coastal sages, and in South Africa as renosterbos (Margaris 1981).

Mediterranean-type ecosystems appear to be highly heterogeneous in the course of time and spatially fragmented, which and this is considered highly characteristic. Periodic heterogeneity stems from interannual, seasonal and diurnal sources of variability in temperature and humidity. Figure 1.2 shows the results of a periodogram analysis applied to data concerning temperature and precipitation combined in Eberger's pluviometric quotient. Data were gathered over a period of 40 years in the city of Thessaloniki. The graph displays a clear-cut peak at a frequency of 0.25, meaning interannual periodicity of 4 years. Analogous interannual patterns have also been reported for most other Mediterranean regions (e.g. Petanidou 1991). In addition to regular interannual climatic fluctuations, broad seasonal fluctuations in climatic variables is another characteristic of Mediterranean climates. For example, seasonal fluctuations of temperature exceeding 55 °C coupled with water content variations in the range of 10-60% are commonly recorded in the upper soil layers. Quite frequently, the seasonal variation in air temperature exceeds 55 °C, and the dry period persists for more than half a year. Figure 1.3 shows the ombrothermic graph for an asphodel desert in Thessaly. Air temperature varied from 0 to 50 °C, while the dry period lasted for 6 months in the first year and 8 months in the second. Pronounced oscillations in solar radiation

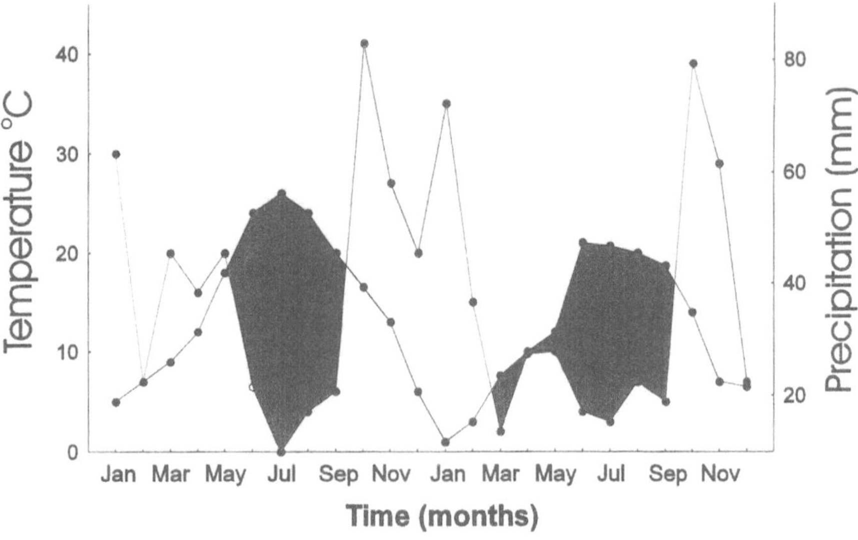

Fig. 1.3. Standard ombrothermic diagram for a Thessaly asphodel semi-desert. (Data from Pantis 1987)

have also been recorded in Mediterranean areas (Lamotte and Bladin 1989). Finally, in addition to long-term interannual and mid-term seasonal fluctuations, strong short-term diurnal variations have also been recorded (Lamotte and Bladin 1989; Stamou and Iatrou 1993).

Interannual, seasonal and diurnal oscillations in climatic variables constitute the more or less predictable component of the Mediterranean climate. However, random catastrophic events may sometimes be of decisive importance. For example, Lamott and Bladin (1989) mentioned the significance of sudden and heavy rainfall for the degradation of soils. Pantis (1987) studied the effect of rainfall intensity on soil erosion in a Greek asphodel desert. He reported that the quantity of surface soil loss is linearly correlated with precipitation lasting for more than 6 h, while it is inversely correlated with plant cover. This author reported that conditions favouring erosion occurred during the period of unexpected rainfall between June and August.

To provide an overall idea of the climatic regime and the soil properties of Mediterranean-type ecosystems, data from eight different Greek sites covered with phryganic vegetation are summarised in Table 1.1. The picture is to some extent representative, although remarkable differences have been recorded in different Mediterranean regions of the world, and some caution is warranted with respect to generalised conclusions. In all sites summer rainfall is very low compared to precipitation in winter. Water deficit during sum-

Table 1.1. Climatic and soil characteristics of eight Greek phryganic formations (Diamanto-poulos et al. 1994)

Characteristics	Sites of Aegean Islands				Sites of continental areas			
Climatic parameters								
Mean annual air temperature (°C)	19.1	18.7	18.1	18.3	18.2	17.6	17.6	17.1
Mean winter air temperature (°C)	13.5	12.8	12.7	12.0	10.6	10.3	10.7	10.9
Mean summer air temperature (°C)	25.1	25.2	23.9	25.4	26.9	25.6	25.0	25.7
Annual rainfall (mm)	502	656	397	586	416	638	1181	475
Winter rainfall (mm)	273	340	215	311	172	31	541	147
Summer rainfall (mm)	4	9	7	9	23	14	27	53
Water deficit during summer (mm)	320	289	322	283	387	260	237	323
Duration of water deficit (days)	150	90	180	120	120	90	90	120
Aridity index	48.8	37.2	56.6	41.3	57.0	38.0	31.2	48.1
Soil parameters								
Sand (%)	35	14	49	35	31	14	45	56
Silt (%)	17	14	14	20	37	24	24	16
Clay (%)	48	72	37	45	32	62	31	28
Loss on ignition (%)	10.2	11.3	8.2	14.7	12.7	20.7	11.4	12.4
Soil moisture content (%)	3.1	2.9	2.6	2.9	2.8	3.2	3.3	2.0
Available water capacity (%)	60	69	62	64	55	40	80	58
pH (1:1 soil:water)	7.6	6.8	8.0	7.6	7.9	7.5	7.0	8.1
Conductivity (μmhos)	800	933	162	110	300	120	102	154
Total N (%)	0.21	0.21	0.28	0.34	0.43	0.34	0.39	0.36
Organic C (%)	2.1	2.0	1.4	1.0	3.1	1.6	1.7	1.6
C/N	10	10	5	3	7	5	4	5
Organic matter (%)	3.67	3.45	2.70	1.79	5.30	2.82	2.83	2.47
Extractable P	0.020	0.006	0.043	0.011	0.050	0.016	0.024	0.048

mer ranges from 237 to 387 mm and lasts 3 to 6 months. It is also worth noting that soils are comparatively rich in nitrogen, poor in organic matter, and almost deprived of extractable phosphorus.

In general, the soils in Mediterranean areas are shallow and the soil profile distorted. Outcropping parent rock usually forms massive blocks. For instance, in a maquis ecosystem on Naxos island, outcropping parent rock covers about 10-20% of soil surface, while surface rocks cover some 20-40% more (Matsakis et al. 1992). The existence of surface rocks in degraded Mediterranean areas is very important for smoothing erosion phenomena. In addition, soil fauna protected beneath surface rocks initiates recolonisation of burnt areas of Mediterranean vegetation (Sgardelis et al. 1995). The depth of the organic layers ranges from 0-5 cm, while the depth of the mineral rocky soils is 0.5 to 1 m (Fig. 1.4). The patchy humus layer is thin (between 0 and 5 cm), and the rock content of the soil profiles is high, usually exceeding 60%. Active carbonate is abundant throughout or in parts of the soil profile, and

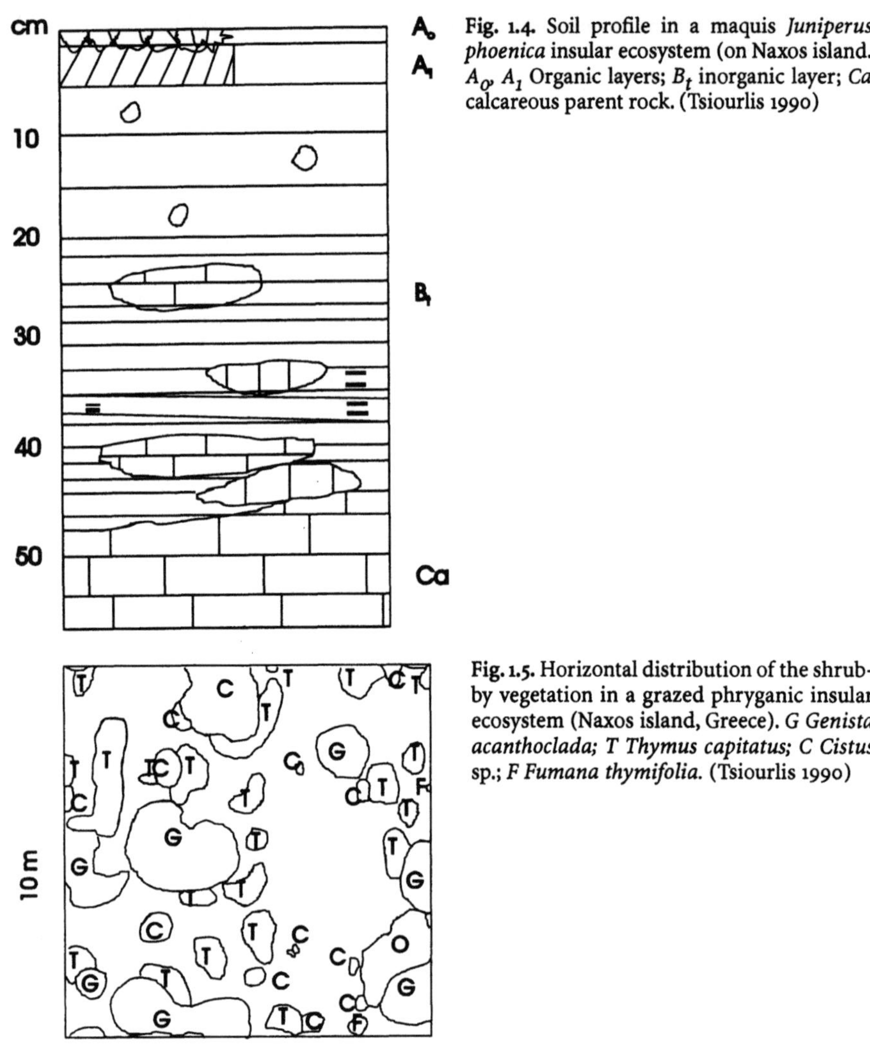

Fig. 1.4. Soil profile in a maquis *Juniperus phoenica* insular ecosystem (on Naxos island. A_0, A_1 Organic layers; B_t inorganic layer; *Ca* calcareous parent rock. (Tsiourlis 1990)

Fig. 1.5. Horizontal distribution of the shrubby vegetation in a grazed phryganic insular ecosystem (Naxos island, Greece). *G Genista acanthoclada; T Thymus capitatus; C Cistus sp.; F Fumana thymifolia.* (Tsiourlis 1990)

iron oxides are also present for most of the year. Finally, in addition to soil erosion, lateral leaching is also characteristic of Mediterranean regions (Stamatiadis and Dindal 1990).

1.2.2
Vegetation

The height of woody plants usually depends on grazing pressure. For example, in areas dominated by *Quercus coccifera*, the height of shrubs in ungrazed

Table 1.2. Number of woody species (No), Margalef's index of abundance (R1), Simpson's index of species diversity (p), and Hill's index of equitability (E) recorded in 14 Greek maquis formations. (Mardiris 1992).

Site	N_o	R_1	p	E
1	10	1.32	0.15	0.87
2	10	1.26	0.13	0.89
3	14	1.84	0.13	0.81
4	11	1.45	0.12	0.87
5	10	1.29	0.23	0.73
6	11	1.40	0.16	0.80
7	10	1.27	0.20	0.78
9	8	1.04	0.21	0.81
10	12	1.59	0.11	0.88
11	8	1.08	0.18	0.84
12	10	1.31	0.26	0.71
13	11	1.42	0.22	0.73
14	8	0.98	0.35	0.65

areas is 1 to maximally 2 m, in modestly gazed areas 0.5-1 m, yet in overgrazed areas up to 0.5 m. Vegetation coverage is also controlled by grazing pressure. Depending on grazing pressure, vegetation appears aggregated, and shrubs form more or less packed clumps interspaced by herbaceous vegetation and dominated by grasses (Fig. 1.5). Irrespective of grazing pressure, species diversity of herbaceous vegetation is high. For example, 60 species were recorded in the maquis of Naxos (Tsiourlis 1990) and 29 in Hortiatis (Iatrou 1989).

1.2.3
Geographical Distribution

To illustrate the interplay of factors determining the geographical distribution of Mediterranean-type ecosystems, I will focus on two examples, the first provided by Mardiris (1992) and the second by Diamantopoulos et al. (1994). These examples concern the geographical distribution of 14 Greek maquis and nine phryganic ecosystems.

1.2.3.1
Maquis Formations

On a large national scale the structural characteristics of maquis, including species richness, diversity and constancy, as well as the distribution of abundance, are relatively homogeneous (Table 1.2). The number of woody species is relatively high, and no great differences are recorded between different regions. Species diversity is also high, indicating co-dominance of several

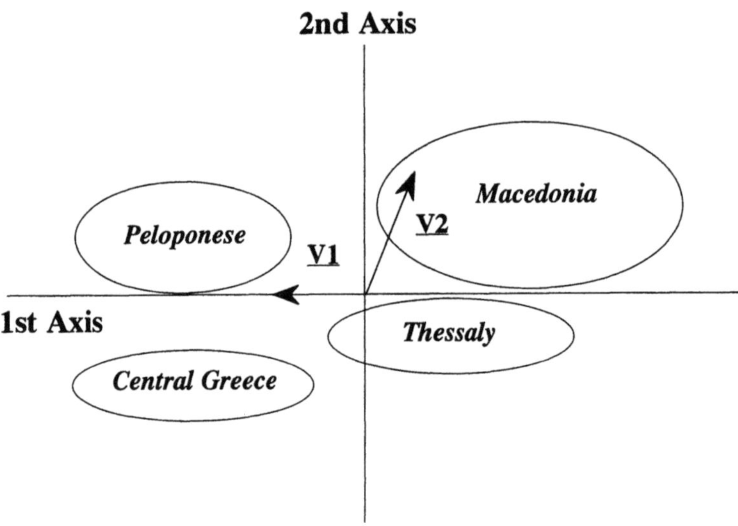

Fig 1.6. Canonical correlation analysis (CCA) ordination of a maquis ecosystem from different Greek regions. V1 Air temperature (annual mean); V2 air relative humididty (annual mean)

species. Moreover, the estimated values of constancy are also high, indicating well-structured communities. The above statements were corroborated by fitting data to the niche pre-emption model of Motomura (Whittaker 1965). It is worth noting that the environmental constant (c) provided by the model is positively correlated with climatic variables. Moreover, in all 14 areas the estimated values for the environmental constant are relatively high, indicating low intraspecific competition as well as successful exploitation of resources. Analogous suggestions have also been reported for other Mediterranean regions (e.g. Whittaker 1972; Naveh and Whittaker 1979; Housard et al. 1980; Bond 1983).

The relationship of abundance to distribution is in agreement with graphical models describing community structure in terms of core and satellite species. Analysis reveals the existence of core species (e.g. *Q. coccifera, Asparagus aqutifolius, Cistus incanus, Philyrea media, Erica arborea, Pistacea lentiscus, Arbutus unedo*) as well as a fair number of satellite species. Slight differences were recorded at a local level (within different sampling sites of the same region) and can be attributed to different core species and/or differences in the dispersion of the satellite species among samples, reflecting local climatic variation.

The distribution of maquis on a national scale is determined by the climatic variables mean annual temperature (*V1*) and mean relative air humidity (*V2*). These variables are the only ones correlated with the distribution of samples along the first axis of a canonical correlation analysis biplot (Fig. 1.6).

Table 1.3. Characteristics of the woody vegetation in eight Greek phryganic formations. (Diamantopoulos et al. 1994)

Characteristics of vegetation	1	2	3	4	5	6	7	8
Winter deciduous (%)	1.1	1.7	0.1	0.1	-	0.3	2.0	1.2
Sesonal dimorphic (%)	90.8	87.1	92.3	80.0	85.0	95.0	95.0	90.7
Succulent (%)	-	-	-	-	3	-	0.2	0.7
Evergreen (%)	2.5	6.7	2.6	8.7	7.0	3.8	0.2	6.2
% individuals with spines	13.3	18.9	8.9	6.3	16.7	7.5	23.4	2.2
% spines in stem	13.3	16.9	7.3	5.3	14.8	7.0	18.6	1.7
% spines in leaves	-	2.0	1.6	1.0	1.9	0.5	4.8	0.5
No. of woody plants m^{-2}	4.4	4.9	6.2	7.9	4.5	3.1	3.2	12.7
% cover of woody plants	62	78	58	54	43	66	41	45
Total plant biomass (g m^{-2})	950	420	899	949	1111	3630	1027	548

1-4, Sites from Aegean islands; 5-8, continental sites

Samples from different geographic regions are ordinated separately. Thus, samples from colder regions are ordinated closer to the right endpoint of the first axis, whilst samples from more arid regions are ordinated closer to the lower endpoint of the second axis.

Anthropogenic factors change the above picture. Species diversity of woody plants is strongly linked to human impact. For example, diversity is high in natural areas (e.g. Mardiris 1992), while in overgrazed areas a few resistant species such as *Q. coccifera* overwhelm the woody vegetation (e.g. Papatheodorou 1996).

1.2.3.2
Phryganic Formations

Mediterranean xeric ecosystems are generally considered to be deterioration stages of the ideal Mediterranean forest (e.g. Horvart et al. 1974; Economidou 1976; Raus 1979a,b; Westman 1981). Moreover, it is often assumed that this degradation process is driven by human activities. Other authors (e.g. Margaris 1981), however, argue on the grounds of ecological convergence in the physiognomy of the world's Mediterranean-type ecosystems and state that phrygana are natural ecosystems well adapted to encountering a severe environment. The fact that the Mediterranean-type ecosystems other than those surrounding the Mediterranean basin, have only recently suffered from intensive human intervention seems to add support to the latter suggestions. However, the discord still persists.

A review of the data (Table 1.3) shows that Greek phryganic ecosystems (representing formations of the Mediterranean basin) display some specific characteristics when compared to ecologically equivalent and convergent formations from California, Chile and South Africa. Greek formations are characterised by an increased proportion of plants with seasonally dimorphic leaves as well as plants with spines, and by a lesser proportion of plants with succulent leaves, while they are the only systems in which plants without leaves as well as species deciduous in winter have been recorded. Diamantopoulos et al. (1994) postulated that these differences might partly be attributed to climatic variations and partly to intensive exploitation (indudinging frequent fires and grazing), which has existed for millennia in the Mediterranean basin. Species diversity of woody plants can also be associated with management history. Comparatively less species diversity has been recorded in the communities of Greek woody plants, this being attributable to the selective pressure of intensive grazing. Apparently, the increased proportion of Labiatae species in Greek phrygana not consumed by grazing animals add support to this assumption.

The relevés from Greek formations were ordinated on a CCA biplot (Fig 1.7) along the first axis in relation to their bioclimatic origin. Samples from the Aegean islands and from continental sites were ordinated separately. Moreover, the ordination of species and relevés along the first axis is correlated primarily with soil variables (loss upon ignition, total-N and C/N ratio), whereas short-term differences in the values of climatic variables are not distinguishing factors. It seems that the distribution of phryganic formations on a major national scale is controlled by more stable characteristics such as bioclimatic type and soil properties.

The two sets of Aegean and continental relevés were treated separately. Ordination of the Aegean samples was controlled by variables concerning humidity conditions, whereas ordination of the continental samples was in accordance with their geographical origin. In the latter case, samples from the wettest western sites occupy the left endpoint of the axis, and samples from the more arid eastern sites are ordinated closer to the right endpoint of the axis. Similar biplots were obtained in cases where, instead of species composition, the dependent variable concerned life forms. It seems that within smaller bioclimatic scales, local differences in climatic variables appear so pronounced as to determine the ordination of samples on the CCA biplots.

The overall conclusion drawn by Diamantopoulos et al. (1994) is that long- and short-term climatic effects, soil properties and anthropogenic factors are synergetically, although not equally, involved in the establishment and appearance of xeric Mediterranean ecosystems. In some areas these formations correspond to climax stages, while in others they are man-made.

1.2.4
The Resilience of Mediterranean-Type Ecosystems

The spatial heterogeneity of Mediterranean habitats originates from the heterogeneity of the Mediterranean landscape as well as from management practices involving overgrazing of the land coupled with frequent fires and the

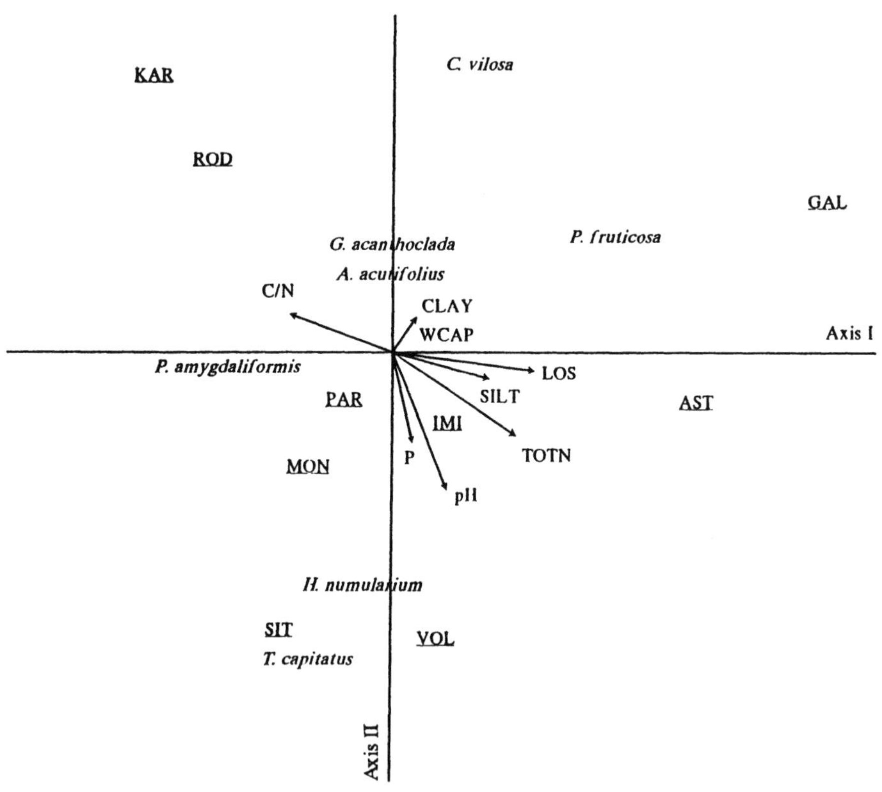

Fig. 1.7. Canonical correlation analysis (CCA) ordination of nine phryganic sites, the dominant plant species and environmental variables. Underlined capitals are abbreviations of sites. Environmental variables: *LOS* loss on ignition; *SILT* silt content; *TOTN* total nitrogen; *P* total phosphorus; *CLAY* clay content; *WCAP* water holding capacity; *C/N* carbon to nitrogen ratio. (Diamantopoulos et al. 1994)

removal of wood. The severity of the environment in Mediterranean-type formations results from the vast temporal fluctuations in climatic variables as well as the fragmented structure of the habitat. The combined result of physical and human impacts is the continuous degradation of Mediterranean landscapes.

Mediterranean vegetation is well adapted to face environmental severity (e.g. Stamou and Pantis 1995) and Mediterranean-type ecosystems appear highly resilient. Fox and Fox (1986) examined the resilience of Mediterranean-type ecosystems in relation to both natural and human-imposed constraints. They concluded that communities show high resilience, notably with respect to succession in the fields discussed and after fires. The authors also tested the convergence hypothesis. That hypothesis is that, due to common climatic and anthropogenic effects, Mediterranean-type ecosystems, although distantly developed, show remarkable similarities in their physiognomy, structure and function. Their overall conclusion is that attributes of

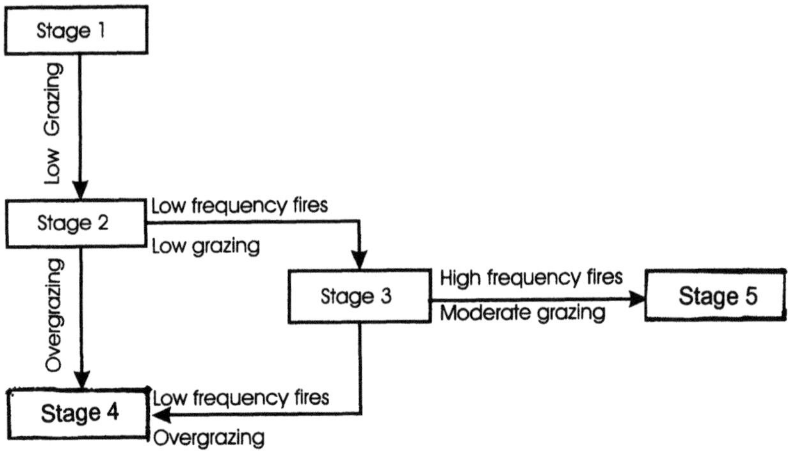

Fig. 1.8. Degradation process of maquis ecosystems induced by grazing and fire. Shaded blocks stand for the final degradation stages (Stamou and Pantis 1995)

Table 1.4. Vegetational characteristics of different degradation stages of a maquis formation. (Stamou and Pantis 1995)

Structural characteristics of vegetation	Stage 1	Stage 2	Stage 3	Stage 4	Stage 5
Cover of woody plants (%)	71	58	45	51	6
Cover of herbaceous plants (%)	19	12	15	24	54
Cover of desirable plants (%)	80	50	4	10	8
Cover of less desirable plants (%)	3	5	6	5	7
Cover of non-desirable plants (%)	7	15	50	60	45
Total plant cover (%)	90	70	60	75	60
No. of woody species	10	10	6	4	3
No. of herbaceous species	15	14	18	22	23
Mean vegetation height (m)	2.5	0.8	0.5	0.3	0.7

fire, and more specifically the frequency of fire, are the most important selective forces to condition the responses of vegetation to other human effects. Finally, they reformulated the drought hypothesis suggesting that summer drought, along with frequent natural disturbances, results in higher resilience in response to human induced effects.

Figure 1.8 shows a qualitative degradation model based on Pantis and Mardiris (1992) and modified slightly by Stamou and Pantis (1995). Model building was based upon the structural characteristics of the communities as well as similarity indices. The structural features of vegetation in different degradation stages are shown in Table 1.4. Three paths of degradation are depicted, all initiated from an undisturbed dense maquis community dominated by *Q. coccifera*. The first path is modulated only by grazing and leads

Table 1.5. Vegetational characteristics induced by fire and grazing

Fire	Grazing
Reduction in vegetation cover	High number of new resprouts confined within
Destruction of the stratification of	Higher rate of biomass increase
Increased resprouting	Decreased litterfall
Activation of germination activity	Increased biomass investment to below ground
Increases in dominance of herbaceous	Nutrient and carbohydrate storage in the below
Increased participation of geophytes	Higher rate of nutrient uptake
	Increased nutrient recycling
	Shorter maturation (flowering) period
	Increased participation of herbs

directly to a community dominated by two typical phryganic species, *Thymus capitatus* and *Balota acetabulosa*. The second path results from the combination of low or moderate grazing pressure and fire. The endpoint of this path corresponds to a typical asphodel semi-desert. Finally, the third degradation path involves grazing and low fire frequency.

As shown above, overgrazing results in phryganic formations, while frequent fire leads to asphodel semi-deserts. The anthropogenic impact on Mediterranean-type formations also induces major changes in microenvironmental conditions such as temperature, humidity and soil characteristics, while changes in the biotic environment are reflected in the physiognomy of the vegetation, species composition and abundance, and in plant activity. Table 1.5 gives some vegetation characteristics of particular significance for the structure and function of the fauna in Mediterranean-type formations.

1.2.4.1
Attributes of Resilience

Beyond the well-documented morphological adaptations of plants (resprouting, depth of the root system, and sclerophyllia or seasonal dimorphism of leaves), attributes of resilience can also be exemplified in the distributional, demographic and physiological characteristics of the plants (Stamou and Pantis 1995). Irregular landscapes and erosion result in small and somewhat larger spatial distribution scales. First-order organisations (with smaller aggregation sizes) are linked to resprouting plants, while second-order organisations (with larger aggregation sizes) are associated with high sexual reproductive effort. The latter enables plants to colonise microsites, thereby allowing for plant establishment. Once established in these microsites, plants resprout and form stable aggregates, thereby reducing the effect of erosion and further improving the local quality of soils. Papatheodorou et al. (1993) correlated the spatial distribution of the sclerophyllous tree *Q. coccifera* with

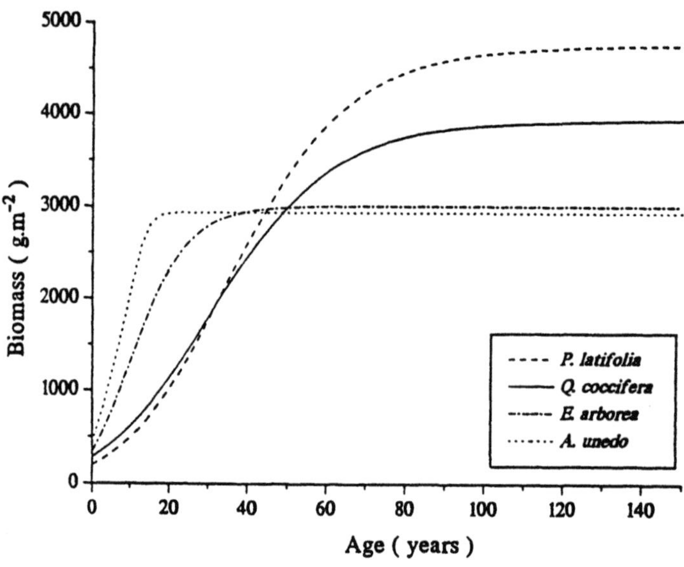

Fig. 1.9. Biomass growth with age in four evergreen-sclerophyllous woody species (Paraskevo-poulos et al. 1994).

the type and intensity of grazing pressure. Grazing changes the age structure of the canopies and forces younger plants to develop in a limited space around the paternal plant. It further results in stronger intraspecific competition among above-ground plant parts. Plants respond by modifying their schedules of biomass allocation, maximising efficiency of nutrient turnover, and increasing growth rates. In general, plant phenology, plant growth, biomass allocation, mobilisation and translocation of biomass and nutrients between different plant parts, nutrient and water uptake, amino acid concentration, etc. synchronise with seasonally oscillating variables (see Arianoutsou-Faraggitaki 1979; Pantis 1987; Papatheodorou 1996, among many others). Stamou and Pantis (1995) suggested that most of the plants dominating deteriorated Mediterranean areas fall into the competitive-ruderal-stress tolerant (C-R-S) class of Grime's adaptive strategies. With regard to growth strategies, woody plants can be classified on a linear continuum. One endpoint of the continuum is occupied by fast growing, high risk, competitor plants, whilst closer to the other endpoint the low growing, low risk, stress tolerant plants are classified (Fig. 1.9). The former are most numerous in areas under low grazing pressure associated with infrequent fire, while the latter dominate in areas subject to intensive exploitation (overgrazing or frequent fires; Paraskevopoulos et al. 1994).

1.2.5
Litter Production and Decomposition

It is generally accepted that the soils in Mediterranean-type ecosystems are nutrient deprived (e.g. Specht and Moll 1983). Litter decomposition as well as the dynamics of nutrient recycling are therefore of crucial importance for their maintenance. The leaves of evergreen shrubs are small, hard and rigid, containing tannins in their epidermis, while at the onset of the dry period, the softer winter leaves of phryganic plants are replaced by smaller leaves rich in sap content. In general, fallen litters are rich in non-easily decomposing materials such as lignins and tannins (Read and Mitchell 1983; Iatrou 1989). General information relevant to the composition of litter is supplied by Mooney (1977), while specific information is provided by Maggs and Pearson (1977a), Mitchell et al. (1986), Tsiourlis (1990), and Papatheodorou (1996). Dead leaves are most numerous among the contents of litter (over 60%), followed by dead wood (about 10%), while the proportion of fruits is almost negligible.

Litter production is relatively high in relation to the total above-ground biomass. In maquis it is continuous, displaying annual or 6 month periodicity (Margaris 1976; Lamotte and Bladin 1989; Tsiourlis 1990; Argyropoulou et al. 1993). In some extreme Mediterranean environments, though, litter production is confined to short periods (Stamou et al. 1994). According to Tsiourlis (1990), a main peak of litterfall occurs in late spring/early summer and characterises Mediterranean-type formations. In many cases the litterfall rate has been correlated with soil water content, litter water content, evapotranspiration and photoperiodism (Specht and Rayson 1957; Maggs and Pearson 1977a; Nilsen and Muller 1981; Radea 1989).

Reabsorption of nutrients such as N and P before litterfall is considered to be an adaptation of Mediterranean plants growing in nutrient limited soils (Maggs and Pearson 1977b; Fouseki and Margaris 1981; Read and Mitchell 1983; Papatheodorou 1996) resulting in an increasing C/N ratio. Increased C/N ratio coupled with the increased lignin content of falling leaves results in a reduced decomposition rate (Stamatiadis and Dindal 1990). In general, annual decomposition rates range from 10 to 40% (Lossaint and Rapp 1971; Yeilding 1977; Fouseki 1979; Tsiourlis 1990; Argyropoulou et al. 1993). Finally, it is noticeable that decomposition rates and soil microbial activities follow seasonal oscillations in temperature and humidity, thus synchronising with plant growth demands (Fouseki 1979; Stamou et al. 1994).

CHAPTER 2

Water Balance in Mediterranean Arthropods

2.1
Water Relations

Water is essential for life. It is a key element in the function and regulation of terrestrial organisms. The body water content of arthropods usually ranges from 65 to 75% (Hadley 1994). Terrestrial arthropods are highly sensitive to humidity on account of their large surface to volume ratio. High atmospheric humidity favouring absorption of water into the body and reducing transpiration rates, tends to result in higher water balance. The opposite occurs with low atmospheric humidity (Block 1996). Consequently, water conditions induce moisture uptake as well as watersaving adaptations such as reduced cuticular transpiration (Cloudsley-Thompson and Constantinou 1983).

Environmental moisture fluctuates seasonally in Mediterranean-type ecosystems, and water conditions govern the abundance, spatial distribution and activity of terrestrial arthropods to a large extent. When coupled with high temperatures, low atmospheric moisture during the summer results in high saturation deficiencies. In addition, the scarcity of significant rainfall strongly affects soil moisture conditions. Mediterranean surface and soil arthropods evidently undergo seasonal variations in moisture conditions ranging from subaquatic to xeric (Poinsot-Balaguer 1988). Vannier (1978) suggested that taxon specific moisture thresholds determine the survival of arthropods and reported relevant logarithmic water potential values of 4.2 for collembolans and 5 for oribatids. Accordingly, water relationships of Mediterranean arthropods have to be discussed with respect to severe water shortage in summer.

Vannier (1978) described a three-step progressive process of soil desiccation. During the rainy period, wet soil particles result in saturated air in soil crevices, pores and animal burrows. At the beginning of the drought period, water begins to migrate downwards and surface-dwelling animals face water depletion. Moreover, dehydration of the substrate gives rise to patchy microenvironments and creates microgradients from saturated to water-deficient sites. During the third stage, liquid water disappears completely and soil animals have to take up water retained in soil particles. This phenomenon determines the extent to which the water relations of Mediterranean arthropods determine their habitat selection, thereby inducing orthokinetic migration.

Apart from summer drought, Mediterranean arthropods also face a water surplus during the rainy season or after thunderstorms occurring chiefly in late spring/early summer and possibly resulting in pronounced mortality (e.g. Paris 1963). Accordingly, water relations in Mediterranean regions need to be interpreted in relation to water surplus in winter and late spring.

2.2
Components of Water Balance

Water balance in terrestrial arthropods has been studied extensively by Edney (1977) and Hadley (1994). Different avenues of water uptake have been described among arthropods in general. Some species absorb water from moist air, while others absorb water from damp particles through cuticular "water channelling" structures (Cloudsley-Thompson 1975). For a good number of species, food-linked water is the main source of moisture (Cloudsley-Thompson 1975).

In some resistant macroarthropods such as isopods, diplopods etc. that lack waterproof epicuticular wax layers, most of the body water is retained in the cuticle and tissues, while minor quantities are distributed in the digestive and reproductive organs and in the haemolymph (e.g. Warburg 1987b). By contrast, the haemolymph is the major water reservoir in the more tolerant microarthropods (e.g. Verhoef and Li 1983).

Smaller quantities of water are lost through body openings, excrements and respiratory paths, but most of the water lost is transpired through the cuticle (Quinlan and Hadley 1983; Hadley and Quinlan 1984; Hadley 1994). Water loss is temperature and moisture dependent. Generally, it increases with increasing temperature and decreasing ambient humidity.

Some arthropods restrict water loss energetically, whereas others seem to tolerate water privation and low body water content. Animals capable of restricting water loss may survive a loss of up to 30% of their body water. Relevant mechanisms include the epicuticular wax layer of arthropods, conglobation in some isopods and diplopods, and fasting (Edney 1977).

Fasting is the most important mechanism employed by most arthropods, such as the majority of collembolans, against prolonged drought. Furthermore, the connection between obligatory fasting and resistance to moisture loss is considered to be a factor regulating population size (Poinsot-Balaguer 1988). Fasting frequently precedes or follows intensive egg deposition of Mediterranean arthropods. Drought resistant eggs deposited early in summer are able to withstand adversity and hatch soon after the onset of the rains. In general, feeding activity is synchronised with changes in moisture, and intestinal content determines the survival of arthropods under stressful

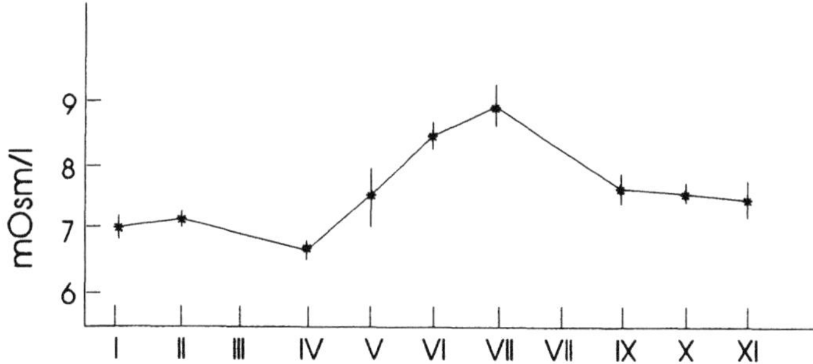

Fig. 2.1. Seasonal changes in haemolymph osmolality of the isopod *Armadillo officinalis*. I-XI months. (Warburg 1987a)

water conditions. Fasting is generally associated with a reduction in high energy-consuming activities, such as locomotion and reproduction, and coupled with diminishing transpiration.

Tolerance strategies involve water absorption directly from moist particles through the ventral tube of Collembola or the uropods of isopods, specific anatomical structures such as the subelytral cavity in some desert beetles (Cloudsley-Thompson 1965; Zachariassen et al. 1987), egg deposition in favourable microsites, behavioural adaptations, and construction of specific structures such as cocoons for egg deposition and larval development (Clouds-ley-Thompson 1982, 1983). Two specific tolerance mechanisms, namely anhydrobiosis and ecomorphosis, involving both structural and metabolic adjustments have been described in only a few collembolans

2.3
The Dynamics of Water Relations

Depending on the concentrations of Na^+ and Cl^-, the osmolality of the haemolymph of arthropods determines their water balance to a great extent. In general, higher values of blood osmotic pressure are measured in summer (Figs 2.1, 2.2), and blood osmolality is also found to be food dependent. Increases in osmotically active compounds, as well as decreases in the amount of haemolymph, result in increased blood osmolality.

Some species such as the collembolan *Orchesella cincta* can regulate osmotic pressure under the progressive dehydration of their substrates, and fasting animals exhibit constant blood osmolality and almost constant body water content in all states (Verhoef and Li 1983). Initially, a rise in feeding activity of the animals is recorded, whereas foraging activity and moulting cease during the advanced stages of dehydration and the animals enter a pre-ecdysial state. Locomotory and metabolic activities as well as transpiration become minimal.

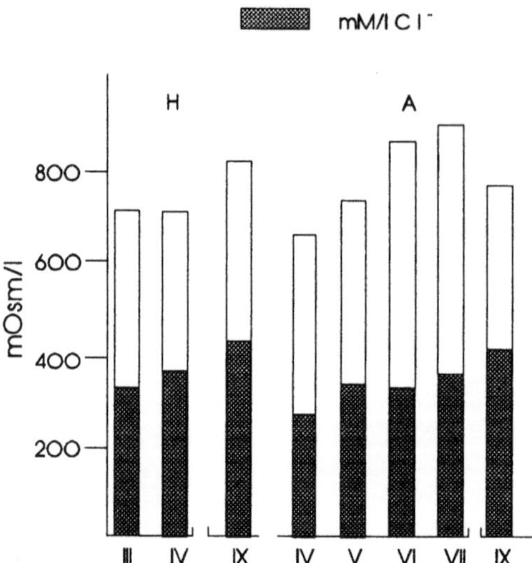

Fig. 2.2. The relationship of haemolymph osmolality to concentration of Cl⁻ in the isopods *Hemilepistus reaumuri(H)* and *Armadillo officinalis (A)* during different months (III-IX). (Warburg 1987a)

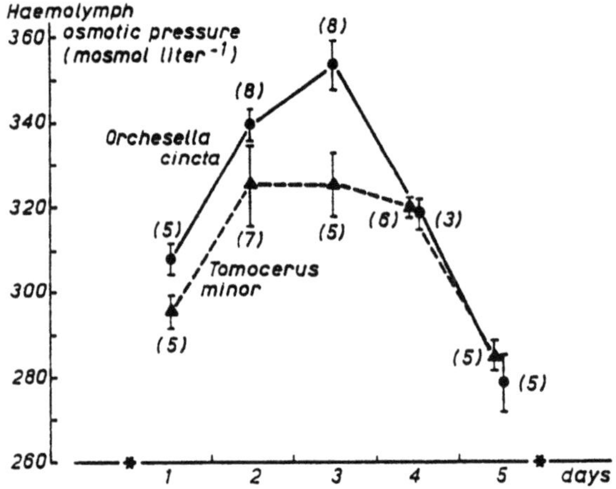

Fig. 2.3. Haemolymph osmolality in an instar of the collembolan *Orchesella cincta* and the collembolan *Tomocerus minor. Numbers in parentheses* are the numbers of test specimens. * Ecdysis. (Verhoef 1981)

In this state, mortality is negligible and animals can survive for long periods. With restored environmental moisture, the animals' activity recovers rapidly.

Other collembolan species, including *Tomocerus minor,* are not capable of osmoregulation. With desiccating substrates – resulting in decreasing food

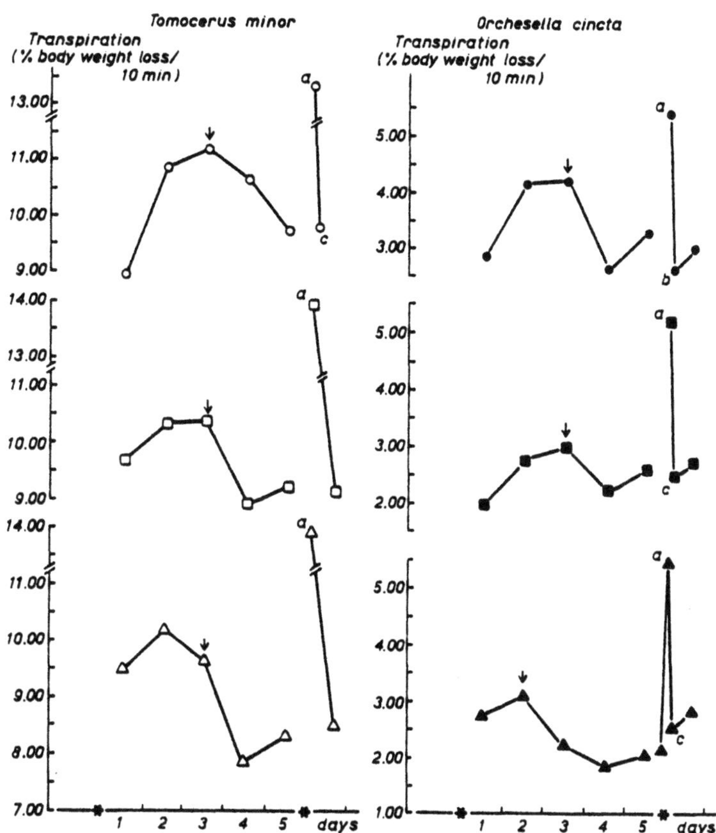

Fig. 2.4. Transpiration during an instar in the collembolans *Orchesella cincta* and *Tomocerus minor*. *Arrows* indicate cessation of feeding activity. * Ecdysis; *a* 10 s after ecdysis; *b* 30 min after ecdysis; *c* 60 min after ecdysis (Verhoef 1981)

moisture – haemolymph osmolality decreases, since a major water source is food. Locomotory activity persists even in progressively desiccating substrates and, after deprivation of nutritional water, the animals consume their own reserves and finally die. Recovery of the population is ensured by a few individuals surviving in favourable microsites.

Fluctuations in blood osmolality during the development of an instar have been observed by Verhoef (1981; Fig. 2.3). Like blood osmolality, integumentary transpiration also fluctuates within instar development. It is high at the end of the feeding period, probably due to abrasion resulting in increasing permeability of the cuticle (Fig. 2.4; Verhoef 1981), and is maximal during ecdysis. Higher transpiration rates recorded during ecdysis emphasise the significance of epicuticular lipids in the water balance of the animals.

Dynamics of water balance may regulate the spatial distribution of collembolans among microsites. Verhoef (1981) stated that the mid-instar rise of blood osmolality pressure in *O. cincta* might be a stimulus forcing animals to seek wet microsites after feeding. Thus, during the feeding period, animals can exploit wider areas. Both the approach to food and its diversity are apparently regulated by haemolymph osmolality, while ecdysis occurs in water saturated microsites.

2.4
Adaptation to Moisture Variations

Water relations are a key element of habitat selection by arthropods. Differences in habitat selection can therefore be partly explained in terms of water budgets. For example, less tolerable lithobiomorph centipedes inhabit rather damp environments, sheltering among less insulated microsites such as the undersides surface stones, fallen litter and bark, whereas more tolerable scolopedromorph centipedes live in more xeric habitats, taking shelter in deep burrows and cracks (Cloudsley-Thompson and Crawford 1970).

Transpiration rate, as well as the resistances of populations of the same species to desiccation appear to be habitat specific. Within the same species, populations from xeric habitats may display lower transpiration rates than populations from temperate ones (Cloudsley-Thompson 1969, 1975). Moreover, considerable differences in transpiration rate may be recorded among different populations of the same species due either to acclimation to moisture conditions within the lifetimes of individuals or to selection or even both (Cloudsley-Thompson 1969).

According to Poinsot-Balaguer and Barra (1991), ancestral adaptations enable Mediterranean arthropods to withstand rapid changes in their moisture regimes. Comparative studies may elucidate the roles of the different components involved in the water balance of arthropods adapting to seasonally varying water regimes. Crawford et al. (1986) conducted a comparative study of the components of water balance in populations of the diplopod *Archispirostreptus tumuliporus judaicus* (=*syriacus*) from both mesic and xeric Mediterranean habitats of Israel and found that the populations' different attributes of water balance depended on the habitat. Thus, desert spirostreptids were more resistant to desiccation than populations of *A. t. judaicus* from mesic habitats. Total water loss, amounting to about 17%, was recorded in the latter after 75-76 h exposure to dry air (<10% relative humidity; Fig. 2.5). In both mesic and xeric populations, total body water content was significantly lower and haemolymph osmolality significantly higher in summer, while in summer more water was retained in the cuticle and tissues than in the haemolymph and gut. These seasonal differences were partly attributed to the lower water content of foods in summer. Baker (1980) found sea-

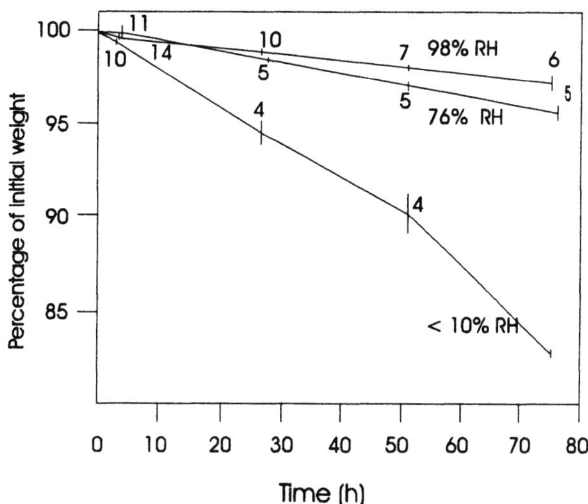

Fig. 2.5. Percentage of water loss in female *Archispirostreptus syriacus* at three different relative air humidities (*RH*). *Numbers* indicate the numbers of test specimens. (Crawford et al. 1986)

sonally varying patterns in the permeability of the cuticle of the diplopod *Ommatoiulus moreletii*. More specifically, he suggested that the new spring and summer cuticles formed after moulting (which occurs in spring) were probably less permeable than the autumnal ones. Furthermore, he linked the impairment of the cuticle, which resulted in increased permeability, to activity and concluded that low summer activity may have resulted in lower cuticle abrasion, while high autumnal activity may have caused it to increase. The overall conclusion is that, irrespective of age and sex, the animals are more resistant under the stress of summer (Ghabbour and Rizk 1979). In addition, Baker (1980) reported seasonal changes in the water content of the animals: the highest values were recorded in spring and summer.

Seasonal variations in body water content differ in species from mesic and desert environments. In the diplopod *A. t. judaicus*, for instance, a higher proportion of body water was recorded in the gut and haemolymph than in the tissues and cuticle during the cool wet period. The reverse was found in the strictly xeric desert diplopod *Orthoporus ornatus*. Intraspecific differences in terms of activity of *A. t. judaicus* from mesic and xeric regions were attributed to differences in habitat (Crawford et al. 1987). In summer, significantly lower body water and haemolymph osmolality were recorded in animals from xeric habitats. These differences were consistent with the properties of the habitats. Desert animals experiencing considerable microclimatic gradients when moving through relatively cool subterranean and hot open surface sites were able to regulate the distribution of water between the gut and gut contents on the one hand and the haemolymph, cuticle and tissues on the other. Shifts in the proportion of water in the gut and gut contents as well as in the haemolymph, cuticle and tissues have been recorded only in populations of desert diplopods; but an analogous osmoregulatory mechanism has also been reported by Warburg (1987b) in certain isopod species from xeric Mediterranean environments.

The comparatively higher transpiration rate recorded in Mediterranean macroarthropods enables them to quickly excrete their water surplus. In addition, changes in the epicuticular lipid layers and membranes of microarthropods affect the permeability of the cuticle and change the transpiration rate. Higher transpiration rates due either to osmoregulation or higher permeability of the cuticle allow arthropods in general to withstand high summer temperatures by thermoregulation via rapid water evaporation through the cuticle. Thus, due to higher transpiration rates, evaporative thermoregulation should be of greater importance in semi-arid arthropods than in desert ones (e.g. Cloudsley-Thompson 1975). Higher transpiration rates may also be beneficial to arthropods living in habitats that suffer from periodic flooding.

In summary, Mediterranean arthropods appear to be moderately resistant to desiccation. Their capability for osmoregulation and the translocation of water among their body parts enable them to overcome summer drought. The higher cuticle permeability recorded in some of these arthropods allows for the excretion of excessive amounts of body water after temporary flooding in winter, late spring and early summer (e.g. Cloudsley-Thompson 1965). Moreover, water relations within ontogenetic stages influence the orthokinetic behaviour of animals, i.e. the distribution of a population among microsites. Thus, mid-instar rise in haemolymph osmolality and transpiration rate as well as ecdysial increase in cuticle permeability can be considered correlates of microhabitat selection.

2.5
Responses of Arthropods to Environmental Extremes

2.5.1
Behavioural Responses

In general, mesic species are less tolerable of environmental extremes such as dessication and cold than species from adverse habitats. However, cautious interpretations should be made under specific circumstances since other factors may equally be involved. For example, Cloudsley-Thompson and Constantinou (1983) compared water relations in spiders from mesic and xeric environments and suggested that body size may be a determining factor. However, among species of the same size, water budgets appear to be habitat and habit dependent. Thus, burrowing mesic mygalomorphs are more vulnerable than desert or arboreal species.

Some arthropods can escape extreme moisture and/or temperature conditions of semi-arid surface soil by burrowing or sheltering under stones during daytime or even for longer periods (Hornung and Warburg 1996). For instance, some spiders from southwestern Western Australia can dig burrows the depth of which depends upon the the spiders aridity of the habitat (Cloudsley-Thompson 1982). The entrances to the holes are usually sealed and create favourable microhabitat conditions within their burrows, living there for several months.

2.5.2
Cryptobiosis

Apart from behavioural adaptations, arthropods display analogous physiological responses to both low temperature and drought that may result in cryptobiosis (Barra et al. 1989). The term "cryptobiosis" is used to describe the reduction of metabolism to a negligible amount. Arthropods enter the cryptobiotic state to survive the extreme conditions that usually occur in unstable environments (Cloudsley-Thompson 1975). Although animals in cryptobiosis can survive extreme conditions for long periods, their viability generally decreases with time (Cloudsley-Thompson 1988).

In general, processes enhancing the resistance of terrestrial arthropods to both cold and desiccation are considered to be complementary (Sømme 1995, 1996), and in most cases desiccation resistant species may be pre-adapted to cold tolerance (Block 1996). For example, the responses of soil collembolans to drought and cold are the same (Fig. 2.6; Poinsot-Balaguer 1990). A phase of active feeding is followed by a fasting period (probably controlled intrinsically) during which morphological and metabolic changes occur. In general, a latent and inactive life allows animals to overcome the environmental severity of Mediterranean habitats, and, apart from structural and physiological adaptations, motionless instars can also survive extremes including desiccated substrates and temperatures as low as -8 °C (Asikidis 1989; Belgnaoui and Barra 1989; Argyropoulou 1993). In fact, by reducing the amount of freezable water, partial dehydration enhances cold hardiness of arthropods. Moreover, desiccation resulting in the accumulation of antifreezable compounds such as polyols lowers the supercooling point of arthropods (Block 1996; Sømme 1996). Two specific cryptobiotic states, namely anhydrobiosis and ecomorphosis have been described in Mediterranean collembolans.

2.5.2.1
Anhydrobiosis

In general, the term "anhydrobiosis" is used to describe phenomena of latent life caused by desiccation (Sømme 1996). As long ago as 1968, Poinsot ana-

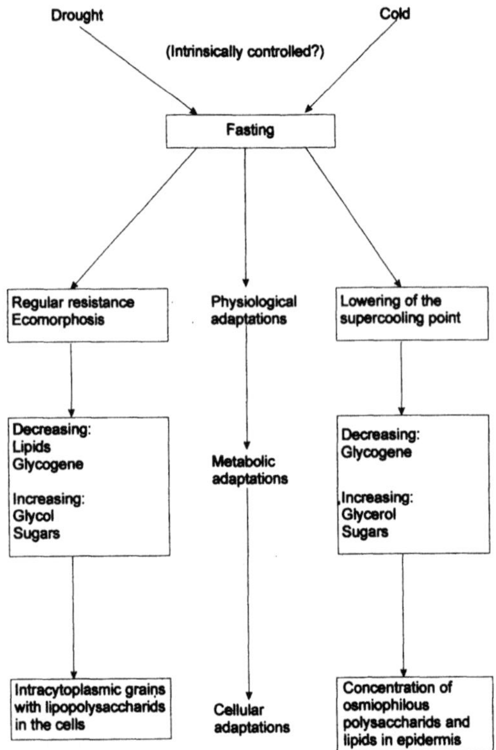

Fig. 2.6. Physiological, metabolic and cellular adaptations of collembolans to cold and drought. (Poinsot-Balaguer 1990)

lysed anhydrobiosis in the collembolan *Subisotoma variabilis.* Anhydrobiosis is induced by progressive desiccation of the substrate, but other external signals and endogenous factors may also be involved (Poinsot-Balaguer 1988). First, animals migrate downwards, remain motionless and lose body water. Changes then occur in their external morphology involving the active contraction and folding of the cuticle and the extrusion of wax, which reduces transpiration. During periods of anhydrobiosis, animals can survive extremely low moisture conditions for several months, and a lowering of the supercooling point to -25 °C, occurs as well. With restored moisture conditions, external morphology revives rapidly and collembolans become active within 1 h. Anhydrobiosis is independent of age and can be entered into several times during the life of an individual.

Aside from morphological changes, Belgnaoui and Barra (1988) also described changes in metabolism in the collembolan *Folsomides angularis.* After 2 months in anhydrobiosis, glycogen which is used for the synthesis of trehalose, has been completely removed from the tissues. Trehalose replaces the water surrounding molecules and membranes, thereby stabilising them under conditions of water depletion (e.g. Sømme 1996). At the same time, lipids are important for the maintenance of metabolism during anhydrobiosis and are used as sources of energy during recovery. After dehydration the glycogen level is rapidly restored.

2.5.2.2
Ecomorphosis

Like "anhydrobiosis", the term "ecomorphosis" is used to describe phenomena involving morphological, anatomical and physiological changes in animals under stress. Ecomorphosis is not linked directly to conditions of water stress in Collembola. Nevertheless, it seems to enable species pre-adapted to cold to adjust their development to the seasonally varying environments that have appeared during the post glacial period (Poinsot-Balaguer 1988).

Like anhydrobiosis, ecomorphosis takes place after downward migration and immobilisation in the organic layers of the soil. It involves morphological modifications of the cuticle and appendices, anatomical changes in the digestive tube, and changes in the metabolism of lipids and glycogen after a period of fasting (Cassagnau 1986). The onset of ecomorphosis is accompanied by voracious foraging which results in an increase in the lipid, glycogen and protein contents of the body. Feeding then ceases and the midgut becomes inactive. A reduction in the lipid and glycogen contents of the body has been recorded at this point, while the amount of protein increases. Initiation of recovery takes place while the animals are still in the deeper soil layers. Bodily reserves are made use of and recovery is followed by upward migration.

CHAPTER 3

Respiratory Metabolism

The introduction of habitat-templet thinking (Southwood 1977; Greenslade 1983) dramatically changed the viewpoint of ecologists with regard to the life history strategies of living organisms. Thereafter, adaptive strategies have been viewed as the outcome of various co-adapted features which can be exemplified not only on demographic but also on physiological and behavioural grounds (e.g. Siepel 1994). In this chapter, the focus is on the respiratory activity of Mediterranean arthropods. Concern with respiratory activity stems from the fact that respiration is the most important path of energy loss. Respirometry can therefore provide a measure of the rate at which animals use their resources to cope with environmental constraints (Peters 1983). The respiratory metabolism of arthropods has been used not only to determine their energy budgets (e.g. Humphreys 1979; Turner 1983) but also to elucidate various aspects of adaptation to cold (e.g. Block 1977, 1979; Sømme 1995), to estimate other parameters of biological activity and demography (Lebrun and van Ruymbeke 1971; Stamou 1986b), and to correlate respiratory activity with changing temperature and humidity (e.g. Testerink 1983). Thus, merging the various aspects of an arthropod's respiratory activity makes it possible to generate important ecological insights into the animal's adaptive strategies.

The complex conditions of temperature and humidity permit trade-offs in the energy allocation of Mediterranean arthropods and hence become the major determinants of their life history tactics. Under strongly fluctuating Mediterranean temperatures, regulation of the dissipation of energy through respiration is apparently of crucial importance for the development of the life cycle of such invertebrates. This can be assessed by determining metabolic characteristics, such as the relationship of respiration to live weight, respiration and temperature, as well as by determining acclimation processes that allow arthropods to withstand the strongly fluctuating thermal regime of Mediterranean regions. Unfortunately, relevant data are scarce. Therefore, to explain aspects of life cycle development in relation to short-term oscillating temperatures the discussion will be based primarily on data presented in three papers by Argyropoulou and Stamou (1993), Stamou and Iatrou (1993), and Stamou et al. (1995)

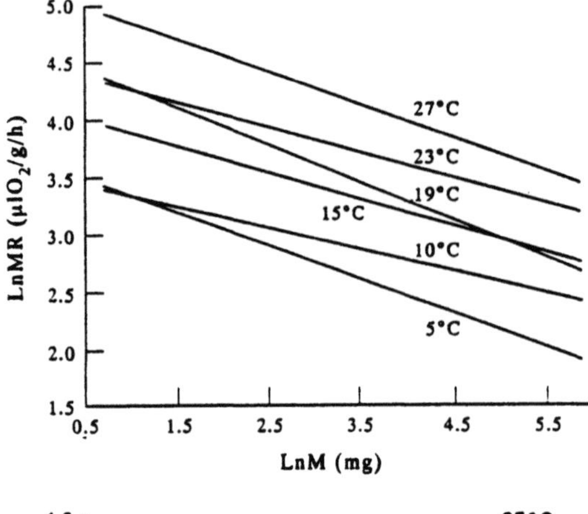

Fig. 3.1. Metabolic rate (*MR*) plotted against live weight (double Ln scale) of the diplopod *Glomeris balcanica* at six different temperatures. (Stamou and Iatrou 1993)

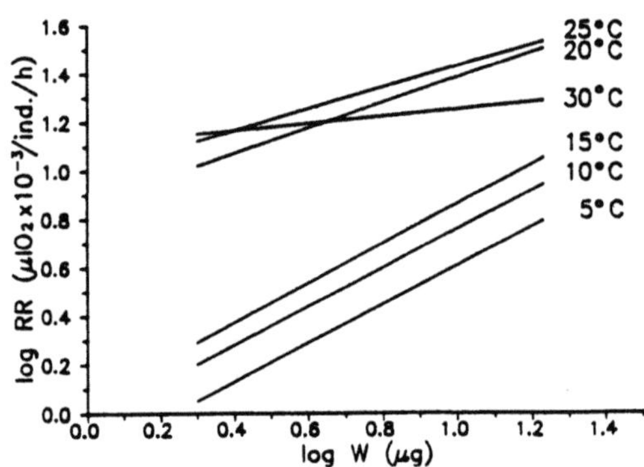

Fig. 3.2. Respiration rate (*RR*) plotted against live weight (*W* double log scale) of the collembolan *Onychiurus meridiatus* at six different temperatures. (Argyropoulou and Stamou 1993)

Animals were collected for respirometry measurements from the evergreensclerophyllous formation at Hortiatis. The animals studied were the long-lived millipede *Glomeris balcanica* (approximately 11 years life expectancy, the univoltine collembolan *Onychiurus meridiatus* (which aestivates in summer and displays density peak in autumn), and the annual oribatid mites *Scheloribates* cf. *latipes*, *Pilogalumna allifera* and *Achipteria oudemansi*. The first-mentioned of the animals peaks in numbers in summer, the second in autumn and spring, and the third in winter.

3.1
Respiration and Weight

To describe the relationship between respiration and live weight, the power function RR=aW^b is used (RR=respiration rate, W=animal's live weight, a=overall mean respiration rate and b=a constant representing the dependence of respiration on live weight). Results have been plotted in Figs. 3.1 and 3.2. In both cases, the regression lines are parallel, indicating the respiration-

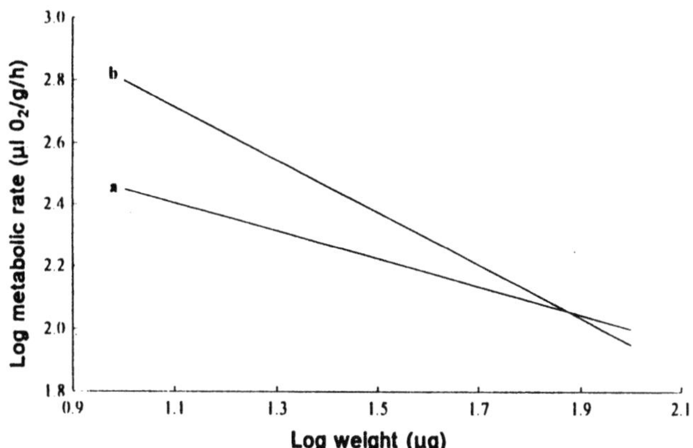

Fig. 3.3. Metabolic rate plotted against live weight (double log scale) of the oribatid *Scheloribates* cf. *latipes*. **a** Estimates made in specimens acclimated to a constant temperature of 28 °C; **b** estimates in specimens during acclimation from fluctuating to constant temperatures. (Data from Asikidis 1989)

mass relationship's independence of temperature, as is the case with the majority of arthropods studied. It is nevetheless worth noting that the values of the parameter b estimated for Mediterranean arthropods fall towards the lower limit of values reported for arthropods from polar, temperate and even tropical regions. Thus, live weight of Mediterranean arthropods affects respiration metabolism only to a relatively moderate extent. Furthermore, the data show that respiratory biomass is more or less evenly distributed among the age classes. This implies that mature and immature individuals are, to some extent, equally efficient with respect to energy transformation. The collembolan *O. meridiatus* constitutes a slight exception in that the response of younger individuals to increased temperature is different from that of older specimens.

Finally, it is evident that although parallel, the regression lines in Fig. 3.2 cannot be replaced by a single one because of differences in their elevations. Further analysis, however, showed no differences in mean respiratory metabolism between two successive temperatures except between 15 and 20 °C. As will be shown below in this chapter, this observation is of adaptive value.

The above conclusions refer to estimates made independently of the thermal regime previously experienced by the arthropods. Nevertheless, when the relationship between respiratory metabolism and live weight of oribatids was studied separately (1) in specimens during their acclimation from fluctuating to constant temperatures and (2) in specimens already acclimated to constant temperature, a temperature-dependent relationship between respiratory activity and live weight was found (Asikidis 1989; Fig. 3.3). The recorded dependence of respiration on live weight was lower during acclimation than afterwards. Moreover, in specimens acclimated to fluctuating as well as specimens

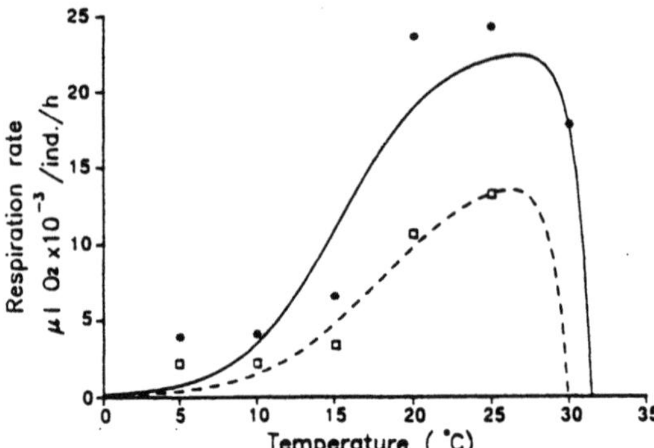

Fig. 3.4. The dependence of respiration rate on temperature in immature *(dashed line)* and mature *(solid line) Onychiurus meridiatus.* (Argyropoulou and Stamou 1993).

acclimated to constant temperature, the lower estimates of b were made at the limits of the temperature gradients. Hence, the effect of live weight on respiratory activity was reduced under a changing thermal regime (from fluctuating to constant temperatures) and at temperatures beyond the thermal range that the animals normally experience. Changing temperatures clearly increase the efficiency of energy transformation in Mediterranean arthropods.

3.2
Respiratory Response to Varying Temperature

Left skewed graphs describing the dependence of different physiological or activity parameters on temperature have been provided for arthropods living in other extreme environments. Such data have been reported, for example, for the Antarctic springtails *Cryptopygus antarcticus* and *Parisotoma octooculata* by Block et al. (1990) and Block (1996). However, these authors exploited only the parts of their experimental data fitting exponential curves.

Data regarding the dependence of Mediterranean collembolan and oribatid metabolism on temperature were exploited by fitting them to the model proposed by Logan et al. (1976; Figs. 3.4, 3.5):

$$RR = a\left[\left(1 + ke^{-pT}\right)^{-1} - e^{(T-T_M)}\right]$$

where RR = respiration rate, T = temperature in °C, p = rate of metabolism increase up to the optimal temperature T_{opt} (i.e. the temperature at which maximum respiration is recorded), T_M = lethal temperature, a = maximum respiration rate, and k = a constant. Furthermore, the following ecologically meaningful parameters can be estimated: the optimal temperature (T_{opt}), the inflection point of the curve (T_{infl}), and the critical temperature (T_{crt}). Finally, the optimum temperature range (OTR) within which changes in respiration rate fluctuate slightly (±0.09) is arbitrarily defined.

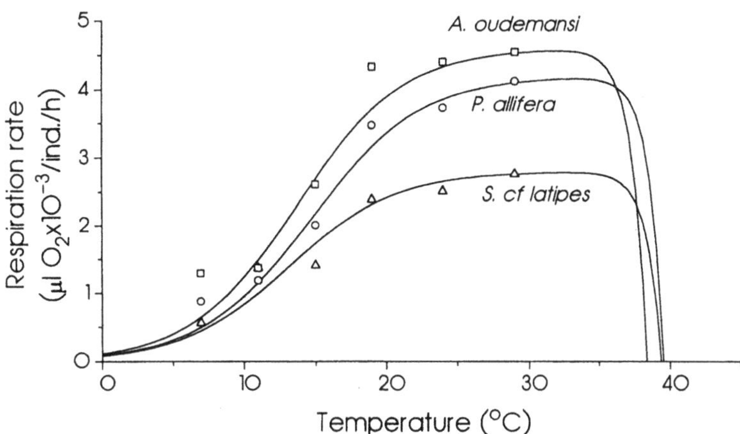

Fig. 3.5. The dependence of respiration rate on temperature in three adult oribatids. (Stamou et al. 1995)

Apart from minor differences, temperature dependence of the oxygen consumption of all animals studied displays major areas of similarity. The dependence of respiratory activity on temperature is skewed to the left. Critical thermal points, up to which respiratory activity remains insensitive to changing temperature, lie within 8 and 13 °C. Below T_{crt} mortality is zero, no development or metamorphosis into subsequent instars occurs, and the animals hardly move. The temperature threshold for egg production lies above this critical point (data from laboratory cultures). It seems that at lower temperatures energy is invested almost exclusively in maintenance. Between T_{crt} and T_{infl} (the latter varying from 13 to 18 °C), respiration rate increases rapidly. Within this range, the effect of temperature on respiration is pronounced. Afterwards, the respiration rate slows down until the optimal temperature has been reached.

It is obvious that animals may balance their respiratory activity within different temperature ranges. Hence, existence is readily inferred of an energy-saving mechanism operating throughout the gradient of temperature tolerance. At lower temperatures this mechanism results in higher energy investment for maintenance, whereas at higher temperatures it ensures higher energy investment in production. From the optimal point up to the lethal threshold, respiratory activity exhibits a precipitous decline. This temperature range is very narrow, indicating that arthropods are highly sensitive beyond T_{opt} temperatures.

It has been demonstrated that warm-adapted arthropods display lower respiratory levels than cold-adapted arthropods (e.g. Block and Young 1978; Dwarakanath et al. 1973; Gromysz-Kalkowska and Tracz 1983; Stamou 1986). This statement does not hold true for Mediterranean species (Table 3.1)

Table 3.1. Respiration rate and Q_{10} values of some arthropods from a Greek evergreen-sclerophyllous formation, estimated over a temperature range of 5 - 30 °C for the diplopod and collenbolan and a range of 5-25 °C for the oribatids

	Respiration rate ($\mathrm{\mu l\ O_2 x10^{-3}\ ind^{-1}\ h^{-1}}$)	Q_{10} value
G. balcanica *(Diplopoda)*	168.32	1.86
O. meridiatus *(Collembola)*	13.34	2.45
S. cf. latipes *(Oribatidae)*	1.83	1.78
P. allifera *(Oribatidae)*	2.50	1.86
A. oudemansi *(Oribatidae)*	3.09	1.69

which display lower metabolic activity than related diplopods (e.g. Byzova 1967; Wooten and Crawford 1974; Penteado and Mendes 1977), oribatids (Wood and Lawton 1973; Luxton 1975; Block 1977; Young 1979; Mitchell 1979) and collembolans (Petersen 1981; Vannier and Verdier 1981) living in either warmer or colder habitats. The same holds for Q_{10} values. Indeed, Q_{10} values estimated for Mediterranean arthropods are low in comparison with those reported for diplopods inhabiting either favourable (Dwarakanath et al. 1973; Gromysz-Kalkowska 1974) or extreme habitats (Wooten and Crawford 1974; Wernick et al. 1983), or for oribatids and collembolans (see references above). Low respiratory activity coupled with low Q_{10} values indicates that the compensatory mechanism mentioned above ensures lower maintenance costs and greater capacity to regulate respiratory activity over a broader temperature range. Thus, Mediterranean arthropods experiencing strongly fluctuating thermal regimes display a greater capacity for balanced respiratory activity over the whole range of tolerance than do arthropods from more stable environments, irrespective of whether these are on average colder or warmer.

The temperature range of 19-27 °C at which minor changes in Q_{10} values were recorded can be considered to be the optimal temperature range for the diplopod *G. balcanica*. Comparatively large optimal thermal ranges are also found among oribatids and collembolans. In general, differences in the dependence of the respiratory metabolism on temperature are attributed either to phenology and duration of activity (Gromysz-Kalkowska 1974) or to the range of habitat selection (Wooten and Crawford 1974). The comparatively low respiratory activity recorded in Mediterranean arthropods coupled with a broader optimal temperature range is compatible with broad habitat selection and longer periods of activity.

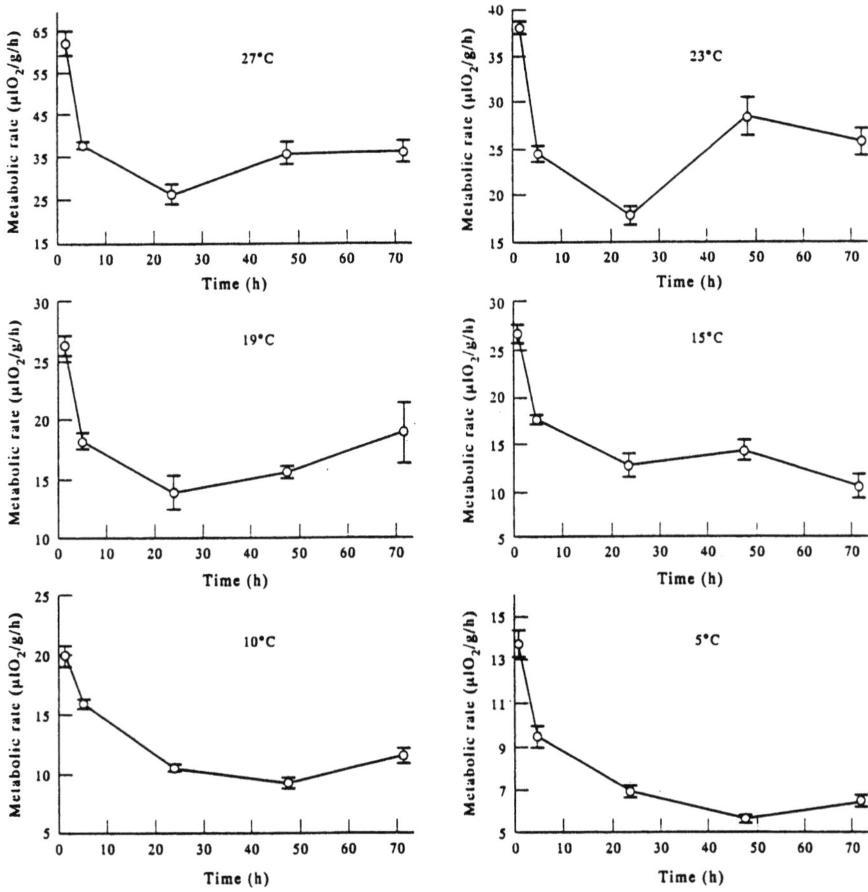

Fig. 3.6. Changes in metabolic rate over time in adult female *Glomeris balcanica* at different temperatures. (Stamou and Iatrou 1993)

3.3
Acclimation to Constant Temperature

In all of the cases discussed above, the respirometre experiments lasted 3 days. Prior to being placed in the respirometre chamber, the specimens were kept in outdoor cultures for about 2 weeks in order to experience local fluctuations of temperature. Consequently, estimates of respiration made during the first hours of each experiment were expected to reflect metabolic activity at fluctuating temperatures, while subsequent measurements corresponded to the respiratory activity of specimens acclimated to constant temperatures. Results for the diplopod *G. balcanica* are depicted in Fig. 3.6. In each individual case a precipitous decline in the animal's metabolism was

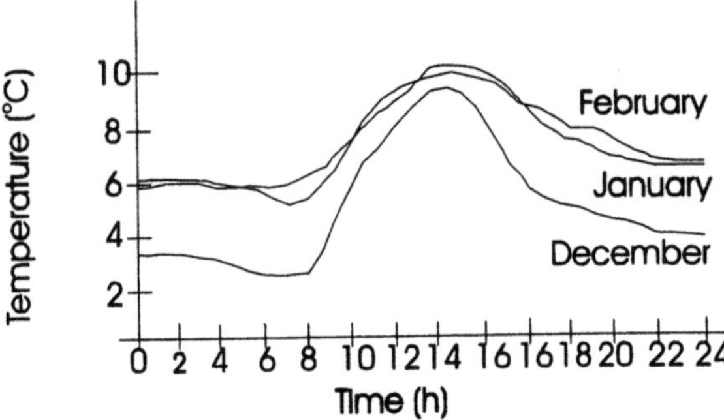

Fig. 3.7. Average diurnal air temperature fluctuations during the period of December-February 1986. From 1 to 6 h temperature remains almost constant (± 1 °C). From 7 to 12 h temperature increases at a rate of 1 °C/h. From 13 to 17 h temperature remains constant (± 0.5 °C). From 18 to 24 h temperature declines at a rate of 0.6 oC/h. (Stamou and Iatrou 1993)

recorded during the first 5 h, whereas after 24 h, the metabolic rate was depressed to 50% of the initial measurement. At lower temperatures (5, 10 and 15 °C) the rate of respiration recorded by the end of the second and third day constantly remained at a low level, while at higher temperatures (19, 23 and 27 °C) it tended to reach a higher level by the end of the second and third day, exhibiting a long-term Precht's (1958) third-type compensation.

It is obvious that the acclimation of arthropods to constant temperatures results in lowering of respiration metabolism. It has often been demonstrated that fluctuating temperatures stimulate arthropod activity (Cloudsley-Thompson 1951, 1953; Dwarakanath 1971a,b; Block and Tilbrook 1977; Stamou 1986b, among others). In the case of the diplopod *G. balcanica,* the response of respiratory activity to a regime of changing temperature is almost immediate. In the light of the daily sequence of alternating constant and changing temperatures (each lasting for 5-7 h; Fig. 3.7), it can be inferred that high costs of maintenance and production at changing temperatures are counterbalanced by energy saved at constant temperatures.

The respiratory behaviour of the collembolan *O. meridiatus* was also more or less similar. The effect of acclimation was studied separately in mature and immature *O. meridiatus* specimens (Fig. 3.8). In mature specimens the metabolism-temperature curve shifted downward without rotating on the second day of the experiment. Thus, in mature specimens no rotation in the curve was recorded since second day recordings were significantly lower than those of the first day over the whole temperature range (Prosser's 1973 translation acclimation-type). On the contrary, in immature specimens a significant difference occurred only at 20 and 25 °C, and a translation coupled with a rotation acclimation-type was exhibited.

The conclusion is that with mature specimens the stimulating effect of fluctuating temperatures holds over the entire temperature range, whereas

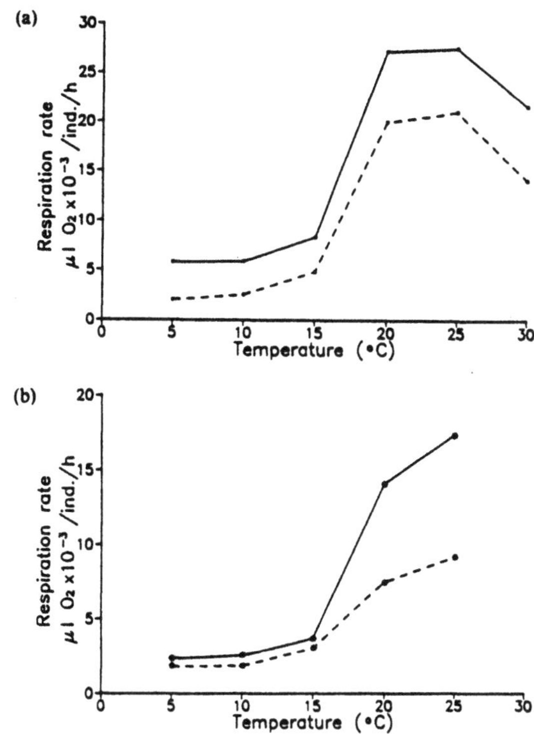

Fig. 3.8a,b. Changes in respiration rate of a mature and b immature specimens of *Onychiurus meridiatus* on the first (*solid line*) and second (*dashed line*) day of the experiment. (Argyropoulou and Stamou 1993)

with immature individuals it is significant only at higher temperatures. Thus, mature specimens are able to profit from even slight but instantaneous increases in winter temperatures, so they can develop as well as produce and deposit small numbers of eggs. In contrast, respiration metabolism is closely correlated with live weight among subsquent immature instars emerging in winter. These appear unable to exploit rises in temperature. Thus, subsequent immature instars cannot develop into adults and dominate numerically during the cold period. When induced by low heat budgets, accumulation of subsequent instars (population reserves) contributes to life cycle synchronisation with the seasonally varying temperatures in Mediterranean regions.

The question as to whether the stimulating effect of increasing temperature is comparable with that of decreasing temperature is of crucial importance in relation to the question whether acclimation offers greater advantages during the cold winter or hot summer. The response of arthropods to sudden increases or decreases in temperature has been studied in oribatids. According to the experimental design, measurements of respiration were made over

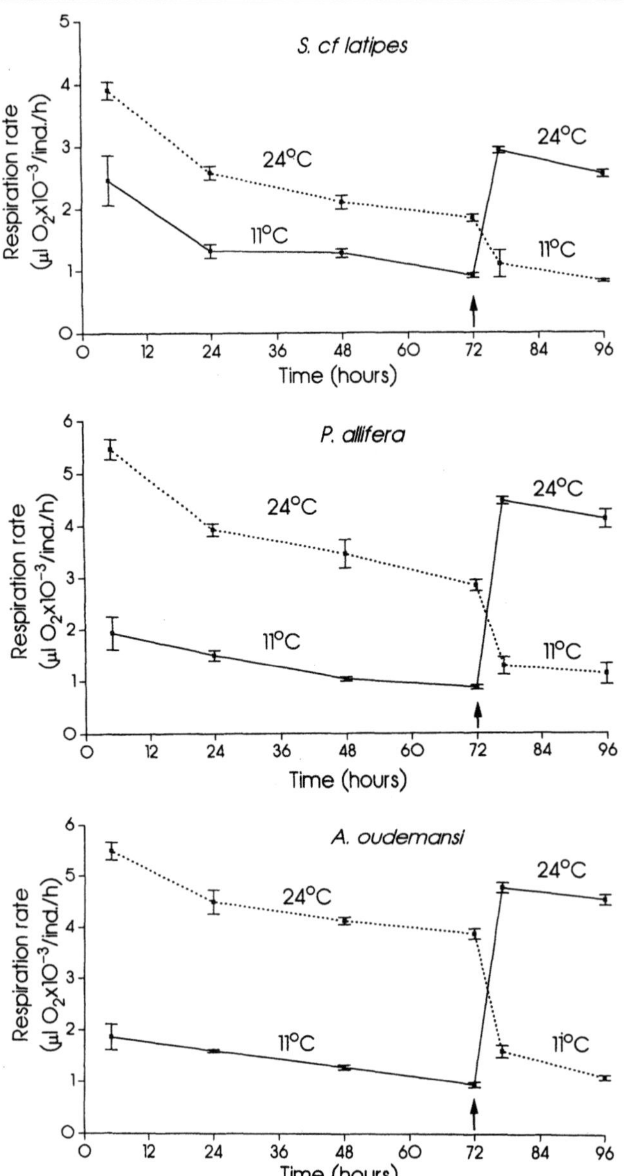

Fig. 3.9a-c. Acclimation processes of the oribatids a *Scheloribates* cf. *latipes,* b *Pilogalumna alli-fera* and c *Achipteria oudemansi* to constant temperature conditions. *Arrows* indicate the moment of temperature change. (Stamou et al. 1995)

3 days at 24 and 11 °C. After 72 h the temperature of 24 °C experienced by the experimental animals was changed to 11 °C, while that of 11 °C experienced

by the second batch of experimental animals was raised to 24 °C. Thus, animals acclimated for 3 days to constant temperature experienced a sudden temperature rise or reduction of 13 °C. Relevant results are graphed in Fig. 3.9. During the three first 3 days of the experiment, the respiratory behaviour of oribatids is similar to that recorded in the diplopod *G. balcanica* and the collembolan *O. meridiatus*. Respiratory depression by the end of the second day was about 30% and a further 10% respiratory decline was recorded after 72 h.

All three oribatid species acclimated for 3 days to 11 °C responded rapidly to rising temperatures (by 400% in the case of *A. oudemansi*). The effect of falling temperature was not as pronounced as that of rising temperature, and respiratory decline ranged from 30 to 60%. It is also worth noting that the overall mean respiration metabolism in animals transferred from 11 to 24 °C was higher than the one recorded in specimens acclimated for 3 days to this temperature. Lowering temperature (from 24 to 11 °C) does not have an analogous effect.

From the above results it can readily be inferred that the response to increasing temperature is a winter adaptation in most animals. At low winter temperatures, arthropods exploit even the slightest temperature increases and are able to accomplish energy-consuming activities such as egg synthesis and deposition. In contrast, the broad optimal temperature range is an adaptation to hot weather. At the high temperatures that occur from late spring to early autumn (when arthropods metamorphose into later developmental stadia and deposit their eggs), metabolic activity is moderate (low Q_{10} value) and, as a result of energy compensation, remains more or less constant over a broad temperature range. Thus, animals can avoid excessive increases in their maintenance cost during the unfavourable period of drought.

To deal with temperature extremes, Mediterranean arthropods have developed respiratory strategies involving low metabolic rates, low Q_{10} values and broad temperature plateaux. Like scorpions from arid regions (Riddle 1978), Mediterranean arthropods conserve metabolic reserves in response to variable temperature conditions. This results in minimising respiratory energy loss. Furthermore, low respiratory Q_{10} values result in metabolic homeostasis following short-term temperature changes. Finally, broader temperature plateaux enable animals to optimise respiratory activity over larger temperature ranges.

The respiratory physiology of Mediterranean arthropods may predictably be modified by thermal acclimation. Hence, it is possible to simulate the respiratory activity of animals under field conditions in order to elucidate various aspects of their life history strategies in strongly fluctuating habitats. Mediterranean arthropods respond rapidly to changing temperatures, irrespective of the conditions of acclimation, and have low maintenance costs at low temperatures, while energy expenditure is comparatively low at higher temperatures. Moreover, the capacity for energy transformation is distributed evenly among the age classes, especially when temperatures fluctuate and animals are active over a broad temperature range. In contrast, rising temperature stimulates energy dissipation, and animals encountering a relatively long period of low temperature in winter can profit from sudden temperature increases to accomplish part of their development. During favourable periods, respiratory activity appears to be relatively temperature independent, and consequently the energetic costs of egg deposition and further development remain low. The overall conclusion is that the Mediterranean arthropods which appear to conserve energy are broad temperature selected, and moderate expenditure of energy resulting from oscillating temperatures is counterbalanced by low energy transformation rates at constant temperatures. This results in broader habitat selection and longer periods of activity. Moreover, Mediterranean arthropods appear highly sensitive to temperatures beyond the optimum. In general, such temperatures in Mediterranean regions are coupled with drought. Accordingly, specific responses are needless, since these conditions are overcome by water adaptations such as anhydrobiosis and fasting (see Chap. 2).

CHAPTER 4

Activity Patterns

4.1
Daily Patterns

The physiological responses to environmental constraints described in previous sections control the activity of arthropods in the field. The daily activity of most animals is controlled by free-running biological clocks, although it is continuously modulated by changes in abiotic variables as well (Cloudsley-Thompson 1982). For example, the activity of some beetle species from Mediterranean pasturelands is determined by an endogenous circadian rhythm as well as by light intensity (Mena et al. 1989). The endogenous rhythm induces a preactive state, while the onset of activity itself is light dependent. Diurnal species (11 out of 19) were active under light intensities beyond 110 lx, crepuscular ones (three species) were active under a light intensity ranging from 110 to 0 lx, whilst four exclusively nocturnal species were also recorded.

The flight activity of crepuscular and nocturnal beetles does not appear to be influenced by temperature or humidity. In contrast, temperature determines the flight activity of diurnal species. In winter and spring maximum activity occurs during the afternoon, whereas in summer some species shift their time of activity towards the morning following a "maxithermal strategy", i.e. they shift their activity to a time at which the maximum sublethal temperature occurs (Mena et al. 1989). Seasonal shifting in daily activity has also been described in diplopods by Crawford et al. (1987). Activity in spring lasts for most of the daylight hours, while in summer the animals are nocturnal. In general, the daily activity patterns of most Mediterranean arthropods follow the climatic variables of the seasons.

Analogous to their desert counterparts, diurnal activities of Mediterranean arthropods can also be conditioned by behavioural thermoregulation such as migration to sites of preferred temperatures or basking in full sunlight etc. (Wooten et al. 1975; Cloudsley-Thompson 1982; Crawford et al. 1987; Hopkin and Read 1992). Thus, differences among diurnal rhythms of different species under different temperature regimes are usually correlated with habitat properties. For example, differences in the activity patterns of the scorpions *Buthotus judaicus* and *Nebo hierochonticus* seem to determine habitat selection. The former inhabits stone terraces and surface rocks, whereas the latter prefers burrows dug under large stones (Warburg and Ben Horin 1979).

In all previous cases, changes in abiotic factors are considered of primary importance for the daily activity of arthropods. However, no general agreement exists on the factors controlling daily activity patterns. In specific cases biotic factors such as avoidance of predators may be equally involved. For example, Cloudsley-Thompson (1981) stated that nocturnality in desert arthropods may be correlated either with avoidance of climatic extremes, or with avoidance of predators or even both phenomena, and analogous explanations might be suggested for the daytime activity of arthropods from Mediterranean habitats.

4.2
Seasonal Patterns

Seasonal activity patterns in Mediterranean areas have frequently been discussed in relation to the climate (Hornung and Warburg 1996). For example, Karamaouna and Geoffroy (1985) reported that surface activity of five diplopod species in an insular Mediterranean formation was confined to the wet period from November to April. Analogous observations were also reported by Baker (1979a) for the millipede *Ommatoiulus moreletii*, in which activity peaks were recorded during the breeding period in autumn/early winter. In summer, most animals displayed little activity – due to their high moisture and moderate temperature preferences – and formed aggregations in favourable microsites (Baker 1980). In the isopod *Schizidium tiberianum* surface activity in early winter is conditioned by the first rains and ceases by late spring. During summer and autumn, animals hide deep in the ground (Warburg et al. 1993). In general, activity peaks in winter characterise arthropods from arid ecosystems, while peaks in autumn and spring characterise fauna from semi-arid ecosystems (Di Castri and Vitali-Di Castri 1981; Magioris and Tsiourlis 1992).

Apart from climatic factors, the effect of grazing and frequent fires can also be recognised in the activity patterns of Mediterranean arthropods. Pantis et al. (1988) analysed pitfall data from an asphodel semi-desert, using the reciprocal average (RA) technique (Fig. 4.1). The seasonal gradient is depicted on the first axis of the graph. Samples from summer and spring are ordinated towards the right end point of the gradient, while samples from autumn and winter are ordinated towards the left side of the graph. Samples from winter and summer occupy the endpoints of the seasonal gradient along with Staphylinidae, Lithobiomorpha (winter samples), scorpions and Scarabeidae (summer samples). Activity peaks of most taxa occur in autumn.

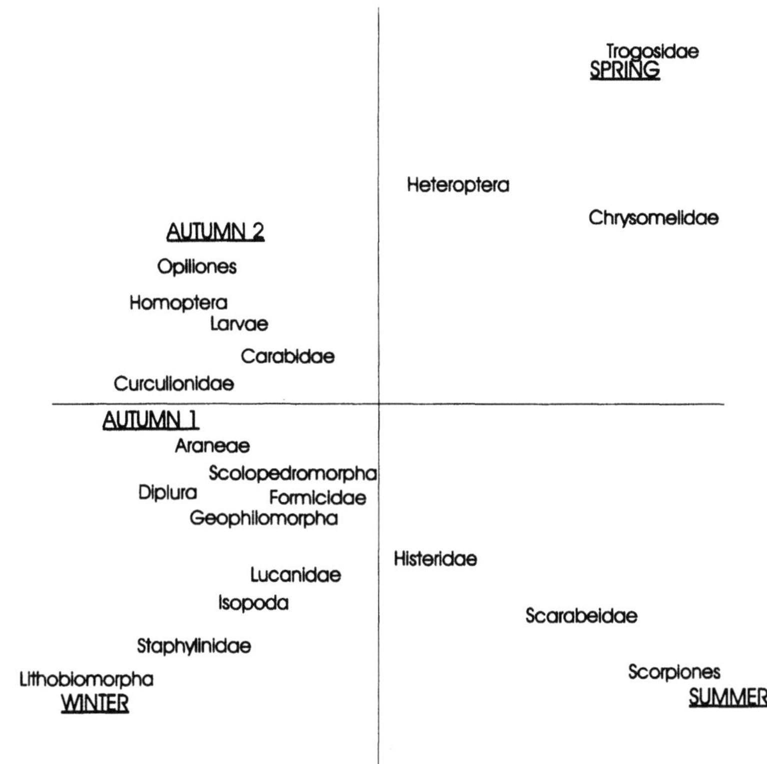

Fig. 4.1. Ordination of some macroarthropod groups trapped in an asphodel semi-desert on the plane of the first two axes of RA analysis. (Data from Pantis et al. 1988)

The activity pattern changes drastically during the transition from summer to autumn and again from winter to spring.

The area studied by Pantis et al. (1988) was subject to intensive grazing. Generally, overgrazing results in a number of microsites maintaining favourable microclimatic conditions as well as in the aggregated distribution of resources. It appears that different taxa, sheltering in favourable microsites, such as Lucanidae, Histeridae, Scarabeidae and scorpions, can remain active under extreme macroclimatic conditions. For example, this is the case with the Mediterranean millipede *Archispirostreptus tumuliporus judaicus*. Crawford et al. (1987) attributed the prolonged dry-season foraging activity of this arthropod as compared with its desert relatives to the large number of sheltering microsites in the vicinity of which the animal can search for its food. It is evident, however, that in Mediterranean regions the importance of sheltering sites varies seasonally (e.g. Hornung and Warburg 1996).

When studying the effects of fire, Pantis et al. (1988) recognised that the dominant taxa, which was active in autumn, was capable of avoiding fires deliberately caused in early summer. The preponderance of highly dispersive Coleptera which on the one hand can escape fire and on the other can rapidly recolonise burnt areas was also attributed to frequent fires.

4.3
Feeding Activity

It is generally assumed that the temporal pattern of feeding activity among terrestrial arthropods parallels variations in climatic variables. However, in Mediterranean regions environmental thresholds determine arthropod activity. Consequently, seasonal changes do not strictly follow relevant changes in the relative humidity and temperature of the atmosphere.

In most cases the water content of the food rather than the relative humidity of the air conditions the feeding activity of soil arthropods (Iatrou 1989). In xeric Mediterranean habitats the water content of food has a direct effect on feeding activity, whereas in mesic habitats the water content of food is high in autumn and spring when feeding activity is at its highest level. Consequently, in the latter habitats temperature principally determines the onset and ending of feeding.

In general, foraging occurs at about 7-25 °C. Feeding activity therefore comes to a halt in winter when the mean monthly temperature is below 7 °C, and it drops drastically in summer when the ambient temperature is above 25 °C. Due to low Q_{10} values, feeding activity also changes slightly within the range of 7-25 °C. Seasonal changes in the feeding activity of most terrestrial arthropods from mesic Mediterranean regions conform with the above scheme (Bertrand et al. 1987; David 1987; Iatrou and Stamou 1989b). For example, except when overwintering and moulting, the diplopod *Glomeris balcanica* is active in Hortiatis throughout the year (Fig. 4.2). Intensive feeding was recorded after cessation of moulting in November and again during the period of oviposition in May/June.

Changes in available water have an overriding effect on the feeding activity of epiphytic arthropods such as pollinating insects. Pollination is of decisive importance for plant reproduction in the more xeric Mediterranean formations. In fact, phryganic plants are mostly insect pollinated (Petanidou and Vokou 1990). The great majority of pollinators are pollen consumers. Nevertheless, in the dominant species of Labiatae pollen is limited, and dilute nectars become the principal reward to pollinators, also representing their main source of water (Johnson and Bond 1992). Thus, the factor determining the activity of insects is the capability of Labiatae to provide both nutritive and water sources, and insects have to synchronise their feeding activity with the flowering period of plants.

In a phryganic formation Petanidou and Vokou (1993), recorded 201 insect species visiting Labiatae, most of them Hymenoptera and Diptera. A remark-

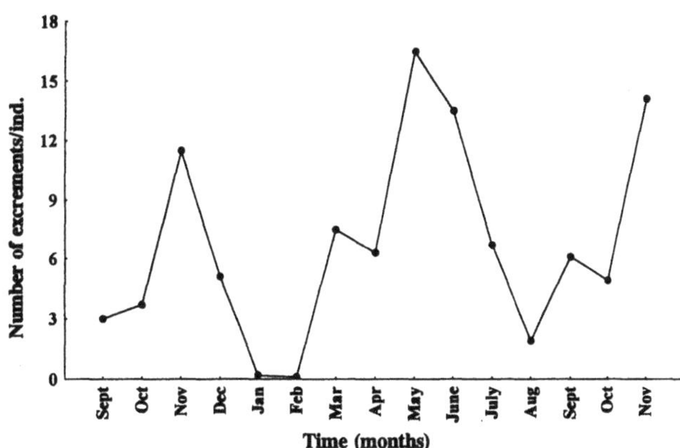

Fig. 4.2. Seasonal changes in the gut content of female *Glomeris balcanica* (Data from Iatrou and Stamou 1989b)

able difference in the composition of pollinator fauna between early and late flowering periods was found by Petanidou (1991). Only 19 out of 201 species were shared by both early and late flowering plants. The early flowering plants were visited by 48 insect species and the late flowering plants by 172. Most insect species exhibited a short period for adult activity lasting approximately 70 days. This restricted them to a few plant species on which they specialized. This phenomenon originated from the fact that plant-pollinator relationships are induced by coevolutionary processes. For example, equipped with good red-perception, the butterfly *Meneris tulbaghia* can rob dilute nectars using its long pliable proboscise. Johnson and Bond (1992) suggest that in at least seven fynbos plants convergence to pollination by *M. tublaghia* occurred independently. The main convergent characters of plants correlate with preferences of insects for large red flowers, dilute nectars, straight narrow nectar tubes and late summer flowering.

In non-pollinating aerial insects herbivory is associated with the quality of foods, which very often varies seasonally. As a matter of fact, insect herbivory of the sclerophyllous species *Banskia oblongifolia* occurs mostly during the developmental phase of the leaves in late spring/summer in conjunction with higher amounts of carbohydrates (Specht and Specht 1992).

Apart from the plant-pollinator relationship, other food relations in aerial Mediterranean arthropods appear mutualistic as well (Lamont 1994). The best example is given in Fig. 4.3 where a food web involving glandular leaves of carnivorous plants (which are best represented in Western Australian ecosystems), sap- and blood-sucking animals, and nectar-feeding arthropods is depicted. Small insects (for the most part nectar-feeding ants) are captured in glandular leaves of carnivorous plants such as *Cephalatus follicularis*, supplementing the plants' mineral supply. In contrast, sap- and blood-sucking

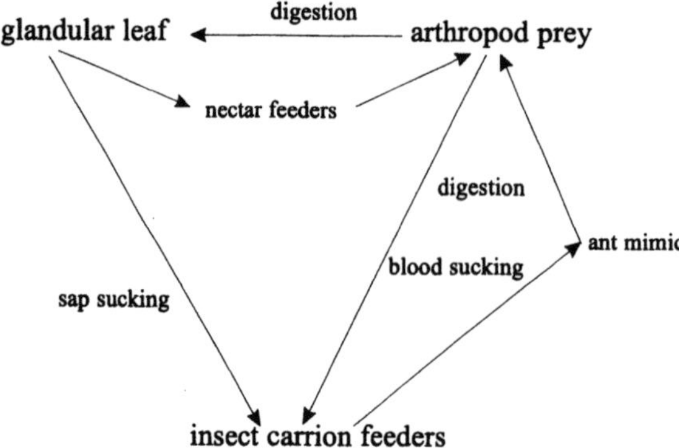

Fig. 4.3. Food relationships of carnivorous plants with arthropod prey and carrion-feeding insects. (Adapted from Lamont 1994)

animals (e.g. the fly *Badisis ambulans*) can profit from captured prey, remains of prey, algae, fungi and bacteria present in the digestive fluids of plants. Apparently, the feeding activity of different arthropod groups involved in such food webs is modulated by the phenophases of the carnivorous plants.

4.4
Oviposition Patterns

In Mediterranean arthropods egg production is usually seasonal, responding to the seasonality of rainfall, temperature and plant phenology. Other factors such as soil conditions and population density may also be involved to a lesser extent (Ghabbour 1983; Ridsdill-Smith 1986). Egg-laying does not strictly parallel the seasons, however, because it is modulated by feeding activity. For example, following feeding activity, egg production in the diplopod *G. balcanica* takes place over the thermal range of 17-25 °C after food and relatively large quantities of foodlinked water have been used for the construction of the egg chambers. Maturation of the ovaries is also dependent upon energy resources accumulated in autumn when the animals feed intensively. As seen in laboratory cultures, egg deposition is accompanied by voracious feeding. The water content of food is evidently of significant importance for both feeding activity and egg deposition. When dried food was offered, both activities almost came to a halt and the majority of animals began moulting (Iatrou 1989).

Temporal changes in the numbers of developing and mature eggs of *G. balcanica* are shown in Fig. 4.4. Developing eggs first appear in December, while mature eggs are first recorded in March/April. In all probability, eggs synthesised in December are never laid. According to Crawford and Warburg (1982),

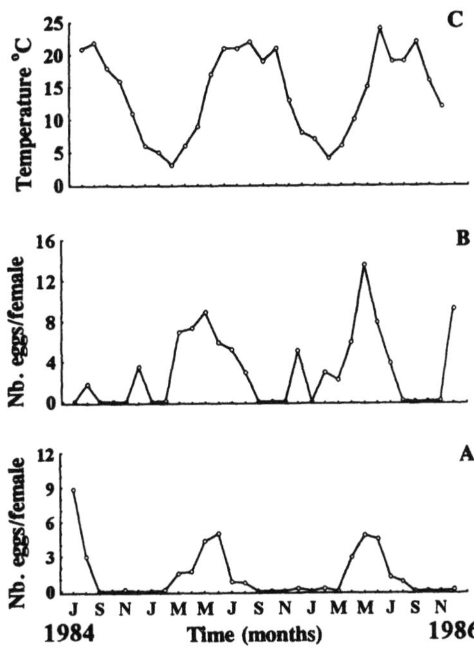

Fig. 4.4A-C. Fluctuation in the number of eggs in the ovaries of *G. balcanica*. A Mature eggs. B Developing eggs. C Fluctuations of the mean monthly temperature in the organic soil layer. (Data from Iatrou and Stamou 1989b)

the absorption of eggs in December is attributable to relative starvation due to the cessation of locomotory activity. In contrast, eggs produced in spring and summer are rapidly deposited from April to mid-July immediately after their maturation. Thus, unlike *Glomeris marginata* from deciduous forests which carries large eggs all year round (Heath et al. 1974), egg synthesis and deposition of *G. balcanica* occur over a short period. Restriction to short periods of activity is very characteristic of arthropod life in Mediterranean regions. For example, the breeding period of the isopod *S. tiberianum* in Mediterranean habitats of Israel lasts 1 month in April (circumscribed breeding), while oogenesis and subsequent vitellogenesis are also brief, lasting 3 months (Warburg et al. 1993).

Iatrou and Stamou (1989b) described the biological activity of *G. balcanica* in Hortiatis as a cyclic sequence of moulting, egg production and egg laying (Fig. 4.5). The onset and cessation of these activities are controlled by temperature and the water content of food. An analogous temporal pattern was also described by Baker (1978b) in *O. moreletii* from south-eastern Australia. The author too stressed the significance of rainfall for the animal's feeding and reproduction. Comparing life histories of Portuguese and Australian populations of *O. moreletii*, Baker (1984) reported that the species

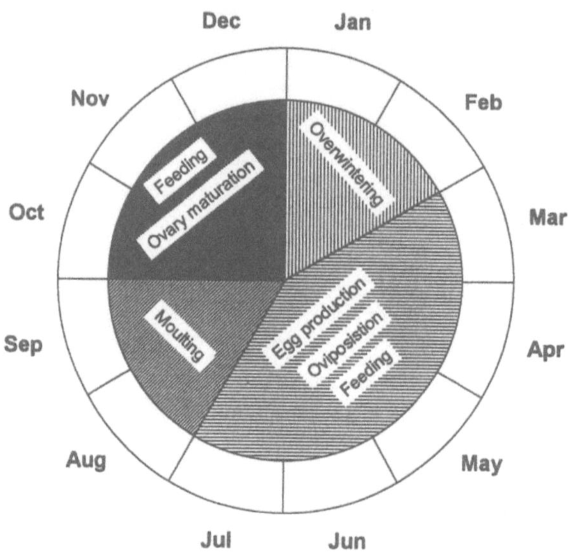

Fig. 4.5. The biological activity of *G. balcanica* in the field. (Modified from Iatrou and Stamou 1989b).

exhibits high plasticity in the timing of its biological activity. For example, breeding schedules – synchronised with local changes in climate – differ in different populations. Nevertheless, data show that in all cases the cyclic sequence of events characterises the life cycle of the animals.

As in other types of ecosystems, the activity patterns of the arthropods in Mediterranean regions are species specific (e.g. David 1995). However, the particular characteristic of Mediterranean regions is that relevant patterns are realised such that a sequence of different activities takes place over short periods. Sequential activity can be viewed as an adaptation to the seasonal components of the Mediterranean climate. By compartmentalising their activity, arthropods perform distinct acts during specified periods when certain environmental conditions more or less prevail. Animals thereby manage to smooth strong yearly oscillations in environmental variables. The next point to be considered is how activity patterns drive the development of life cycles in environments that display strong diurnal, seasonal and interannual fluctuations.

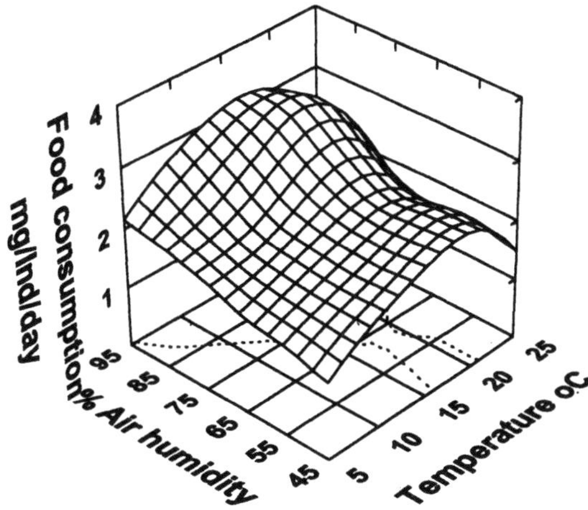

Fig. 4.6. The short-term effect of varying temperature and relative air humidity on adult feeding activity of the diplopod *G. balcanica*. Data were fitted by least squares

4.5
The Joint Effect of Temperature and Humidity on Feeding Activity and Demography

Although factors such as calcium carbonate, salinity, pH value and predation to some extent affect the distribution and activity of Mediterranean arthropods (Ghabbour 1983), temperature and humidity are the principal factors to condition the distribution and dynamics of animals. In general, conclusions regarding the short-term effects of temperature and humidity on arthropods are drawn from studies undertaken independently for each factor. Nevertheless, the effect of temperature depends on the level of humidity, and the joint effect of temperature and humidity on the biological activity of arthropods appears somewhat convoluted. Indeed, temperature and humidity thresholds (above which sharply increasing or decreasing rates are recorded) may result in phenomena that are linearly or quadratically interdependent of one another.

4.5.1
Short-Term Effect of Varying Temperature and Humidity

The effects of temperature and humidity on arthroplods are exemplified by Hortiatis. The graph featuring the temperature and moisture dependence of feeding in the diplopod *G. balcanica* in the laboratory appears rather simple (Fig. 4.6). Feeding activity increases with increasing moisture, and a jump in

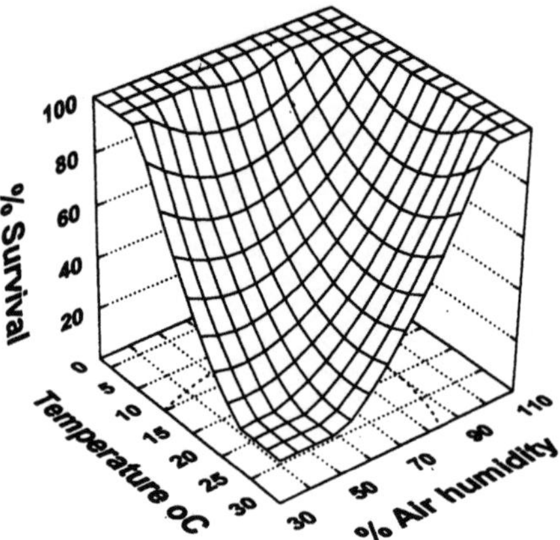

Fig. 4.7. The short-term effect of varying temperature and relative air humidity on adult survivorship of the oribatid *Scheloribates* cf. *latipes*. Data were fitted by least squares.

the curve of the graph is obvious at about 65% relative humidity. The quadratic effect of temperature is highly significant, and the maximum rate of faeces production is recorded at 20 °C for every humidity level. A further increase in temperature up to 25 °C results in decreasing feeding activity, and the animals start moulting. It seems that in general a temperature of 25-27 °C is the upper activity threshold for most Mediterranean arthropods. At higher temperatures, feeding ceases and they enter a latent stage showing enhanced resistance to heat.

Analogous observations concerning feeding activity as well as survivorship in response to varying temperatures and humidity have been made for most oribatid and collembolan species sampled at Hortiatis. The combined short-term effect of temperature and humidity on arthropod survival is shown in Fig. 4.7. At low temperatures survivorship is independent of humidity, while at temperatures beyond 12 °C arthropod survival depends upon the humidity level. In general, different phenomena affecting animal survival, such as cuticular transpiration, are insignificant below certain critical temperatures (Cloudsley-Thompson 1975).

The survival and feeding activity of arthropods commonly takes place above a humidity threshold between 65 and 85% relative humidity. It seems that, like their temperate relatives, the short-term response of the Mediterranean animals to air humidity is of the all-or-none type. The dependence of activity on temperature appears to be more complicated. On the basis of data gathered on Hortiatis it is possible to compile graphs showing the average dependence of arthropod locomotory and/or feeding activity on temperature (Fig. 4.8). Relevant graphs are comparable to those describing the dependence of respiratory metabolism on temperature. Accordingly, graphs of

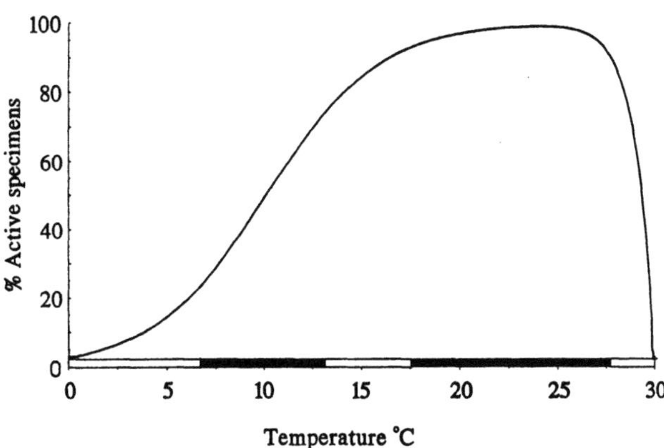

Fig. 4.8. The short-term effect of temperature on arthropod metabolic activity

respiration can be used to illustrate the temperature dependence of activity parameters other than thermoregulation. Activity can take place at temperatures from 5 to 28 °C, although some species such as the oribatid *Scheloribates latipes* and the collembolan *Onychiurus meridiatus* are active at temperatures up to 30 °C. At up to 7 °C activity is low, but the optimum temperature range of 16-27 °C is rapidly attained.

With respect to the temperature coefficient, Lebrun and van Ruymbecke suggested as early as 1971 the Q_{10} values obtained are similar irrespective of the biological parameter on which these estimations are based. It is worth noting that low Q_{10} values estimated for the arthropods from Hortiatis in the range of 5 °C to the optimum temperature using feeding activity or survivorship parameters were in all cases identical to those estimated using respiration metabolism parameters.

The effect of temperature on the activity of Mediterranean arthropods does not generally differ from that on arthropods inhabiting either temperate (Striganova 1972) or desert habitats (Wooten and Crawford 1975; Cloudsley-Thompson 1982). However, the former, which experience milder yet strongly fluctuating thermal regimes, show minor differences with regard to (1) lower Q_{10} values, (2) higher basal and upper critical points, and (3) a broader range of thermal plateaux extending from about 16 to 27 °C.

4.5.2
Energetics

Similarities in feeding and/or locomotory activity in response to temperature and humidity are primarily qualitative, whereas cautious quantification of

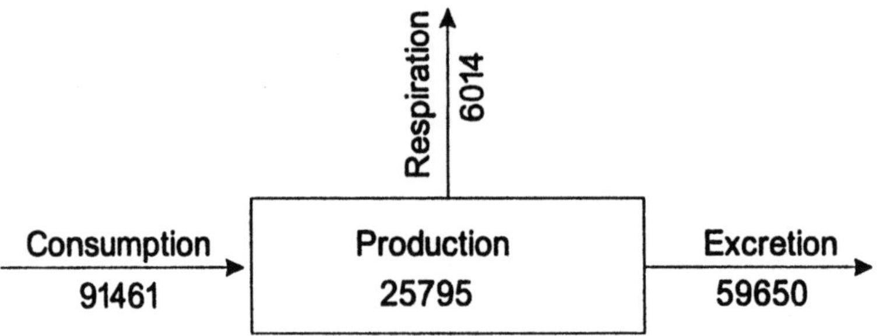

Fig. 4.9. The energy budget (kcal m⁻² y⁻¹) of *G. balcanica*. (Data from Iatrou 1989)

the activity of arthropods should be made in the field. For example, average food consumption of *G. balcanica* in the laboratory is 9.35 mg g⁻¹ day⁻¹, while the corresponding average faeces production is equivalent to 5.03 mg g⁻¹ day⁻¹. Extrapolation of these data allows for the estimation of an average annual consumption of 24.37 g m⁻² in the field. This value corresponds to 5.7% of annual litter production. Of the litter consumed, 53.8% returns to soil through excrement. Although the above picture is indicative of the role of a dominant species in litter decomposition, generalisations appear risky. In fact, much higher (up to 100 g m⁻²) as well as much lower (3 g m⁻²) annual consumption rates have been estimated in related species, and deviating relevant extrapolations have been obtained in other Mediterranean habitats such as the evergreen-sclerophyllous formations of southern France (David 1995).

Macroarthropods, more particularly diplopods such as *G. marginata* or *G. balacanica*, are the principal agents of litter decomposition in Mediterranean ecosystems. Initiating litter breakdown, macroarthropods determine the stability of food webs as well as the rate of litter disappearance. Hence, estimations of energetics could provide useful insights regarding the role of macroarthropods in the decomposition process (again, cautious interpretations should be made). A layout of the energy budget of *G. balcanica* from Hortiatis is illustrated in Fig. 4.9. The low maintenence cost of the animal is evident, since comparatively lower amounts of energy are consumed for respiration demands.

The average caloric content (ash-free) estimated for male animals is 3.9874 cal mg⁻¹, while a significantly higher caloric content of 5.0724 cal mg⁻¹ was estimated for female animals. Moreover, the caloric content of female animals varies significantly with age, and the higher values were recorded in reproducing individuals. No such differences were recorded in male animals. Finally, the estimated respiration and assimilation coefficients are R/P=0.23 and A/C=46.2%, respectively.

Parameters of food consumption of Mediterranean arthropods are in general agreement with findings from most world habitats. The notable exception is the assimilation coefficient recorded in *G. balcanica* which is higher than that estimated for desert (Wooten and Crawford 1975) or temperate counterparts (Mason 1970). Unfortunately, no relevant data for other Mediterranean arthropods exist, but it may generally hold true that low maintenance cost of arthropods is coupled with efficient exploitation of food resources.

4.5.3
Long-Term Effect of Temperature

The long-term effect of temperature is mainly impressed on demographic parameters. Hence, demographic responses to changing temperatures are the main determinants of life history strategies in Mediterranean arthropods. So far, life history schedules have been reported for arthropods from ephemeral, constant or adverse habitats, and relevant tables depicting r, K and A life history categories have been compiled. Life histories of arthropods from seasonal environments hardly fit such patterns. It is expected that strongly fluctuating temperature and humidity regimes in combination with intensive exploitation of the land lead to the development of specific life history schedules compatible with the specificity of the Mediterranean environment. More specifically, demographic characteristics such as the timing of egg deposition, duration of larval development, and arrest in life cycle development (e.g. quiescence, aestivation), allowing for the synchronisation of phenologies with seasonally varying temperature and humidity are of great importance.

To describe life history correlates in response to changing temperature, extrapolation of laboratory data is needed along with actual field data. Successful rearing of most arthropods in the laboratory is possible at temperatures above a certain threshold. For example, oribatids and collembolans from Hortiatis were successfully cultured at temperatures beyond 19 °C (Asikidis 1989; Argyropoulou, unpublished). At intermediate temperatures of 15-17 °C the life cycle development of most species was completed, although no egg deposition were recorded. Below 12 °C no oviposistion and no development into later stadia were recorded, while the animals hardly moved. At temperatures below 5 °C, no mortality was recorded and the animals remained in quiescence preserving their ability to respond immediately to subsequent rising temperature. Early instars develop more slowly at all temperatures than later ones, while intermediate stadia, e.g. protonymphs and deutonymphs of oribatids, develop even faster (the quadratic effect is highly significant; Fig. 4.10). The rate of larval development increases from 17 to 23 °C and decreases slightly in the transition range of 23-27 °C.

Higher mortality of earlier life stages is frequently reported for arthropods of Mediterranean provenance (e.g. Warburg et al. 1984; Bercovitz and Warburg 1985; Crawford et al. 1987; Matthiessen and Ridsdill-Smith 1991) and is

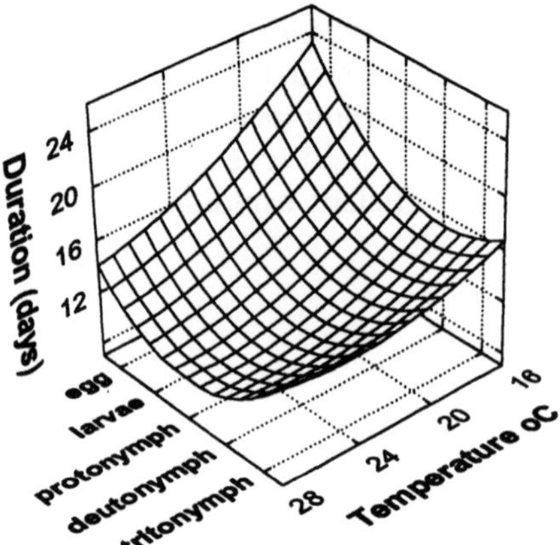

Fig. 4.10. The effect of temperature on the duration of development of the juvenile instars of the oribatid *Scheloribates* cf *latipes.*

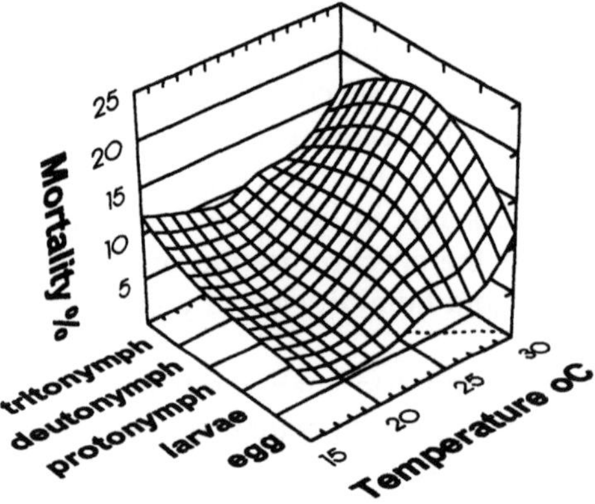

Fig. 4.11. Laboratory survivorship of immature oribatid *S.* cf. *latipes*

attributed to abiotic factors such as flooding, entrapment on the soil surface during the hot dry period or even predation. Data from Hortiatis show that higher ontogenetic cost is counterbalanced by lower reproductive mortality. This statement putting the emphasis on evolutionary history contrasts with explanations paying more tribute to the effect of environmental constraints. Graphs depicting the survival of oribatid and collembolan species from Hortiatis in the laboratory are comparable (Fig. 4.11). Survival of immature animals is low, indicating high ontogenetic cost, and less than 50% of the eggs released manage to attain maturity. In contrast, in almost all cases of survival of mature animals, the curve shows a sigmoid decline (Fig. 4.12) and the mortality plateau (within 8 and 20 weeks) corresponds foremost to those

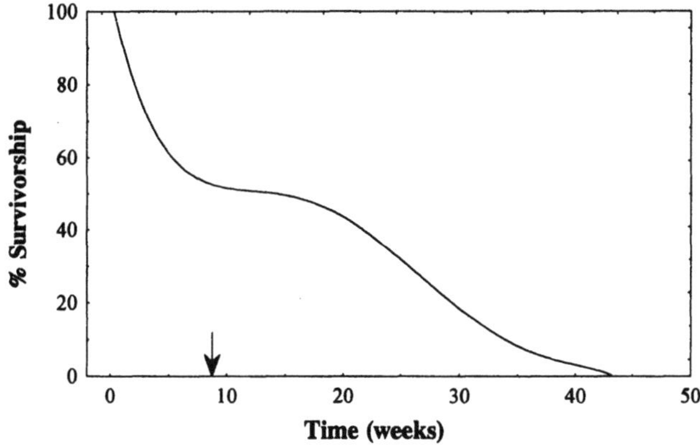

Fig. 4.12. Laboratory survivorship of the oribatid *Pilogalumna allifera*. The *arrow* indicates the onset of maturity. Data were fitted by least squares

stages that represent the highest potential reproductive values. In Mediterranean arthropods, adaptive emphasis is put on adult longevity accompanied by low reproductive mortality at the expense of the viability of the immature stages.

4.5.4
The Effect of Thermal Past on Demographic Parameters

It has been shown that in some poikilotherms from both arid and temperate regions the response to temperature depends on the immediate thermal past (Cloudsley-Thompson 1953; Prosser 1973; Newell et al. 1974, among others). As long ago as 1951, Cloudsley-Thompson stressed the importance of fluctuating temperatures for the successful maintenance of two species of tropical diplopods in the laboratory, reporting that constant temperatures result in a lowering of their activity probably due to long-term thermoregulation. In the case of Mediterranean arthropods, an energy-saving compensatory mechanism was revealed (Chap. 3) controlling the short-term metabolic response of animals to varying temperatures. The question is to what extent does the operation of such a mechanism affect demographic parameters resulting in long-term thermoregulation? Or, to put it in other words, to what extent is the demography of arthropods affected by the animals' thermal past? The response to this question is of crucial importance for extrapolating laboratory data to field conditions (Stamou 1986b).

Short-term exposure of Mediterranean arthropods to constant temperatures usually stimulates oviposition. In contrast, long-term maintenance at constant temperatures results in irreversibly depressing egg deposition. Thus, short-term metabolic thermoregulation stimulates recruitment to popula-

Table 4.1. Experimental design for studying the effect of thermal past on the oviposition and moulting schedules of the dipolopod *G. balcanica*

Experiment	Thermal past	Experimental conditions
1	Acclimated to field temperature regime	Constant laboratory conditions
2	Acclimated to field conditions	Fluctuating (16-29 °C) laboratory conditions
3	Acclimated to field temperature regime	Fluctuating field temperature regime (10-25 °C)
4	Stocked for 1 year at constant temperature conditions (22 ± 2 °C)	Fluctuating field temperature regime (10-25 °C)

tion, while the longterm operation of this mechanism entails lower oviposition rates as well as delays in the oviposition-moulting sequence.

The above statements are exemplified by *G. balcanica* sampled on Hortiatis. A layout of the experimental design is shown in Table 4.1. Reproductive effort is more or less evenly distributed among age classes of mature individuals, and the oviposition period in the field lasts from April to July, followed by the moulting period lasting from July to October (Fig. 4.13). The same oviposition-moulting sequence is also followed by animals cultured in laboratory, while there are pronounced differences in the duration as well as intensity of the phenomena. Specimens from the field (experiment 3) exhibit longer oviposition periods and deposit a relatively large number of eggs compared with animals from laboratory cultures. Animals acclimated for 1 year to constant laboratory conditions (experiment 4) as well as others maintained at fluctuating temperatures in the laboratory (experiment 2) show a shorter oviposition period, while the onset of oviposistion is delayed by 1 month. These individuals display a significantly lower oviposition rate (0.16 eggs/female per day) than do animals acclimated to field conditions (varying from 0.31 to 0.34 eggs/female per day in the different experimental series). The highest oviposition rate of 0.48 eggs/female per day was recorded in animals acclimated to field conditions during their acclimation to constant temperatures (short-term thermoregulation).

Data concerning non-metabolic thermoregulation in Mediterranean arthropods are generally missing. However, behavioural regulation of temperature analogous to that exhibited by desert counterparts (basking, horizontal and vertical migration, e.g. Cloudsley-Thompson 1975) is likely to occur. Moreover, acclimation to high temperatures (increasing the ability to withstand lethal temperatures) cannot be excluded. In general, lethal temperature may vary in relation to the thermal past, while habitat properties, body size and relative air humidity may also be involved (Cloudsley-Thompson 1988).

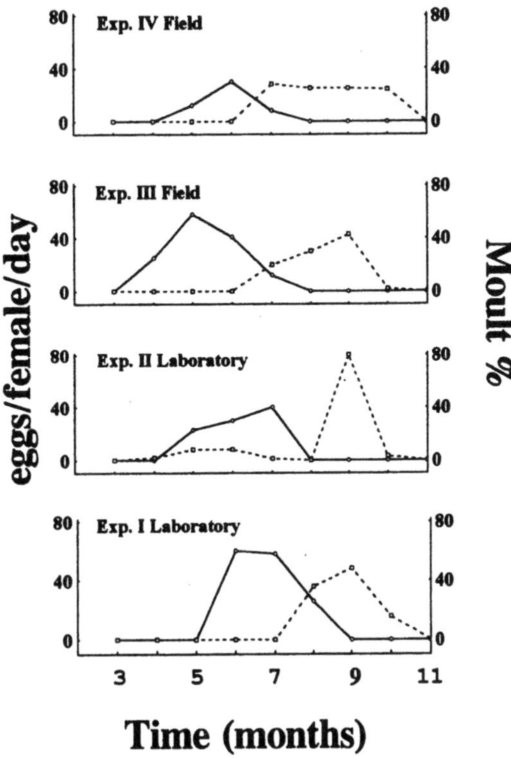

Fig. 4.13. Mean monthly oviposition rate and percentage of moulting in four different experimental series (I-IV) in *G. balcanica*. Oviposition rate is represented by a *solid line* and percentage of moulting specimens by *a dashed line*. (Data from Iatrou 1989)

CHAPTER 5

Life Cycle Tactics and Development

As will be shown in the following sections, the fauna of Mediterranean eco-systems is mostly cosmopolitan. Hence. matters concerning plesiotypic life histories (i.e. ancestrally dictated life history patterns) can be investigated in Mediterranean regions. Consideration of life cycle patterns allows for elaboration of the concept of life cycles complying with the seasonality of the Mediterranean climate. The seasonally varying Mediterranean environment by no means implies uniform demographic schedules, as was presumably thought by David (1995). On the contrary, since it involves variables oscillating over large ranges, seasonality forces the pluralistic expression of demographies in accordance with physiological constraints and the evolutionary history of the fauna. The interplay between the need to conform with environmental seasonality on the one hand and historical constraints on the other, determines the way in which patterns are realised in the field. That is the reason why some promising species, such as the white-fringed weevil *Graphognathus leucoloma* fail to be successful in Mediterranean environments. Matthiessen (1991) claims that this arthropod is poorly adapted to Mediterranean regions probably due to its high demands as to food quality and moisture during reproduction in early summer in conjunction with its relatively high temperature and moisture requirements for larval development and adult emergence in summer. It is obvious that under the constraints of the Mediterranean climate *G. leucoloma* is unable to realise the patterns dictated by its evolutionary history.

5.1
Life History Characteristics

Key features of the highly heterogeneous Mediterranean environments (e.g. seasonality of thermal regime, timing of seasonal precipitation, nature of habitat retreats etc.) appear to have a disproportionate influence on the life cycle patterns of arthropods in specific cases. For example, Juchault et al. (1993) found that environment mainly affects reproductive strategies and reported partly environment-dependent thelygeny in the isopod *Armadilli-*

Table 5.1. Comparison of life history characteristics of the diplopod *A. t. Judaicus* from a mesic and xeric habitat. (Bercovitz and Warburg 1985)

Life history traits	Mesic habitat	Xeric habitat
Larval stages in the first year	3-4	6-7
Age at maturity (years)	8	6
Life span (years)	11	9
Egg wet weight (mg)	5.96±0.40	7.26±0.10
Egg caloric content (cal)	22.12±0.60	25.93±0.91

dium vulgare. In a different habitat, Iatrou (1989) suggested environment-independent reproductive strategies, while he reported environment-dependent oviposition schedules in *Glomeris balcanica.* Apparently, it is difficult to assign key demographic features to Mediterranean arthropods.

Comparative studies of the demography of arthropods from Mediterranean and other habitats are illustrative. Evaluation of relevant data indicates the prominent variability of Mediterranean environmental gradients, showing that differences within semi-arid Mediterranean habitats may be more pronounced than differences among temperate and subhumid Mediterranean habitats. Indeed, intraspecific comparisons (Bercovitz and Warburg 1985, 1988; Crawford et al. 1987) reveal that pronounced differences in the appearance of life history traits have been recorded in populations of the millipede *Archispirostreptus tumuliporus judaicus* from different semi-arid habitats, in addition to negligible differences in morphometric characteristics. In xeric habitats, animals grow faster, reach maturity at a younger age, have a shorter life span and a higher energy investment in eggs (Table 5.1). Clutch size is higher in xeric environments, while pellet production is delayed in mesic habitats for about 3 months. Development within egg pellets and postemergence development are faster in xeric than in mesic habitats and adults attain maturity 2 years earlier. The age at maturity is 6 years in xeric and 8 years in mesic habitats. Finally, the life span in the xeric environment is 9 years compared with 11 years in mesic habitats. The authors viewed these differences as a mechanism to counterbalance environmental adversity. Larval mortality in more hostile xeric habitats is expected to be higher, and shortening the larval stage may save the animals from dying. These differences can also be considered in terms of heat budgets. In fact, lower heat budgets may result in a longer life span, longer generation time and lower investment in eggs in mesic habitats where lower average temperatures are recorded. Different populations also differ with regard to their developmental schedules. This is probably due to differences in the occurrence and duration of favourable seasons. In mesic environments moulting occurs during the dormant period in winter, while in xeric habitats it occurs after dormancy. Moreover, hibernation is shorter in xeric environments than in mesic ones.

In contrast, the other hand, interspecific comparison of the life cycle development of the millipedes *Glomeris marginata* from a temperate habitat (data

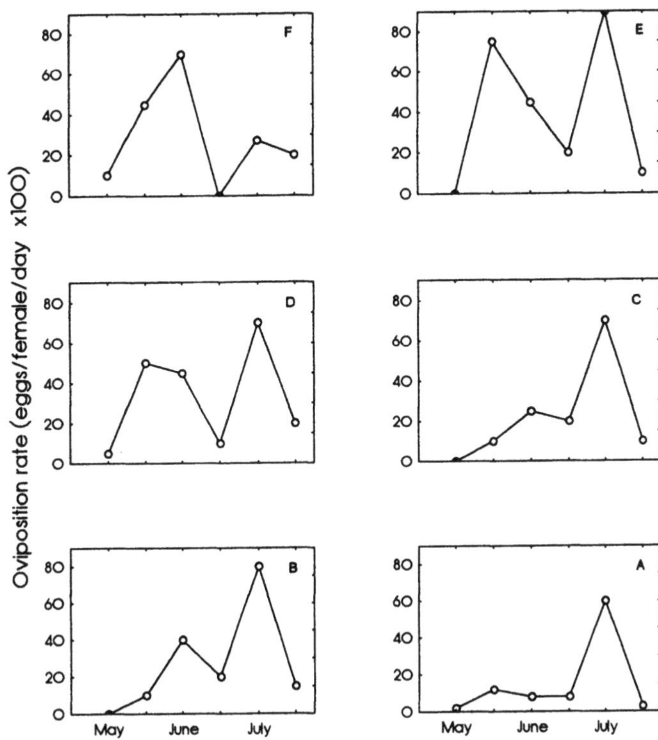

Fig. 5.1. Fluctuations of the oviposition rate in six different weight (age classes) of the diplopod *Glomeris balcanica*. *A* 70-120 mg; *B* 140-160 mg; *C* 160-180 mg; *D* 180-220 mg; *E* 220-260 mg; *F* 260-310 mg. (Data from Iatrou 1989)

from Heath et al. 1974) and *G. balcanica* from a subhumid Mediterranean one (data from Iatrou 1989; Iatrou and Stamou 1990) revealed vast areas of similarity. Both species display comparable fecundity (0.34 eggs/female per day in the former and 0.27 eggs/female per day in the latter). In both species reproductive effort is similarly distributed among mature age classes, and oviposition schedules are analogous. The oviposition period for all mature age classes lasts from early May to late July (Fig. 5.1). Younger animals deposit most of their eggs at the end of the oviposition period, while older ones lay eggs at the beginning of this period and intermediate age classes display bimodal egg deposition. Life cycle development in both species is comparable.

5.1.1
Life Span

In comparison with temperate species, most Mediterranean arthropods show shorter life spans. Short life span is generally considered to be an adaptation to climatic constraints (e.g. Poinsot-Balaguer 1988; Ghabbour 1983). Wallwork (1982) described a short life span among woodlice of arid and semi-arid regions, while populations of the same species may enjoy shorter lives in xeric than in mesic habitats. An analogous observation was also reported by Crawford et al. (1987) among millipede species. It seems that a short life span agrees well with the strongly fluctuating Mediterranean environment. Nonetheless, in accordance with Norton (1994) it can hardly be considered as an apotypic trait of adaptive value. Presumably, under mild temperature conditions a short life span may originate from relatively high heat budgets that allow for a relatively rapid life cycle.

5.1.2
Age at Maturity

Due to higher heat budgets than in temperate relatives and lower maintenance costs than in arid counterparts encountering harsher environments, the age at maturity of Mediterranean arthropods is likely to be short. Evidence shows that this rule holds true in many cases, while many exceptions have also been reported. For example, maturation lasts for less than one third of the life span of different Mediterranean arthropods from Hortiatis (Iatrou and Stamou 1990; Stamou et al. 1993a) and from south eastern Australia (estimates based on data by Baker 1978b) living in comparable environments. In contrast, maturation takes up about two thirds of the life span of the millipede *A. t. judaicus* and the coleopteron *G. leucoloma* (among many others) living in more severe environments. Moreover, different times leading to maturity have also been reported (e.g. Baker 1978a,b; Iatrou and Stamou 1991).

Notwithstanding interspecific discrepancies, the age at maturity of Mediterranean arthropods is to a large extent in tune with the predictable seasonal and interannual cycles. For example, Iatrou and Stamou (1990) concluded that maturation in *G. balcanica* coincides with the approximate 3-year interannual cycle recorded on Hortiatis. Indeed, anamorphosis (up to the fourth free-living stage) takes about a period of 9 months and epimorphosis (up to the tenth free-living stage) about 18 months. Thus, the animals attain maturity after 2.5 years. As in other Mediterranean arthropods (e.g. David 1982), the age at maturity of *G. balcanica* appears to depend on annual and seasonal cycles rather than on the number of moults.

5.1.3
Reproductive Strategies

Reproductive strategies have been discussed in detail in papers by Cole (1954), Gadgil and Bossert (1970), and Charnov and Shaffer (1973) while a critical review has been written by Sterns (1976). In summary, animals having only a short time to maturity and experiencing low premature mortality are semelparus, as would be expected. Semelparity is assumed in temporally predictable habitats in conjunction with uniform spatial distribution of resources. In contrast, iteroparity – that is the distribution of the reproductive effort over time – is the appropriate strategy against either reproductive failure in unfavourable periods (Callow 1978) or heterogeneous spatial distribution of resources (Blower 1969). Distribution of reproduction over time results from distribution of the reproductive effort among the different age classes, from differing time spans to maturity, or even from life cycle timing leading to overlapping age distribution.

As to sexuality, Norton and Palmer (1991) argued that parthenogenetic species are broadly adapted. This stems from the fact that among parthenogenetic species, selection operates on clonal grounds and is usually associated with considerable genetic variation. By contrast, selection in sexual populations, acts on individual loci or linkage groups, so, the chance of broadly adapted whole genomes evolving is greater in parthenogenetics. Norton and Palmer (1991) concluded that broadly adapted thelytokous populations are at an advantage in abiotically unstable habitats where changes in local environmental conditions occur continuously. Obviously, the authors linked the non-stability of habitats favouring parthenogenesis to abiotic unpredictability over time rather than to spatial heterogeneity. The authors also studied reproductive tactics of oribatids, considering biotic rather than abiotic unpredictability to be more important for sexuality. Within a coevolutionary context they argued that in complex communities – inducing higher biotic unpredictability – plastic genomes are a prerequisite for the survival of populations facing sexual parasites, sexual predators or sexual competitors.

In the light of the above assumption, the situation remains controversial in Mediterranean regions characterised by a highly heterogeneous distribution of resources coupled with predictable seasonal and interannual oscillations in climatic variables. Thelytoky and sexual reproduction, semelparity and iteroparity, as well as precocity and differing time spans to maturity seem to be strategies employed with equal probability.

5.1.3.1
Thelytoky and Sexual Reproduction

At first glance, taking into account life history traits such as shorter generation times, relatively low fecundity, and the need for recolonisation of microsites after anthropogenic disasters, the contribution of parthenogenetic species might be important for Mediterranean regions. Nevertheless, as was argued in the previous section, sexuality and parthenogenesis are the by-product of trade-offs among abiotic and biotic predictabilities over time. Sexuality is preferred in abiotically stable environments favouring the establishment of complex communities, whilst parthenogenetics is favoured in abiotically unstable habitats.

The above scheme fits well to Mediterranean arthropods living in habitats with predictable seasonal and interannual fluctuations of environmental variables. It is also evident that in Mediterranean regions climatic predictability is accompanied by hazardous anthropogenic disasters resulting in spatial heterogeneity. Thus, unlike the circumstances of arthropods from extreme environments where parthenogenesis is a key factor, obligatory or facultative thelytocous parthenogenesis or even female dominance are of minor importance in Mediterranean habitats. Likewise, temperate and Mediterranean environments hosting complex communities seem to favour sexual reproduction.

The advantage of sexual over asexual reproduction lies in the re-combination of genetic material that may be more favourable for survival in spatially heterogeneous habitats (e.g. Siepel 1994). In seasonally predictable, spatially heterogeneous environments such as Mediterranean ones, greater genetic variation may ensure establishment of offspring at particular microsites. Thus, sexual reproduction in Mediterranean regions can be viewed as a self-maintaining adaptation to spatially uncertain habitats. As a matter of fact, only three out of 26 oribatid species recorded in the evergreen-sclerophyllous formation on Hortiatis can reproduce parthenogenetically. Moreover, when studying reproductive tactics in oribatids along a huge Mediterranean gradient, Sgardelis (1995) recorded a sex ratio of approximately 1:1 in larger species, while female-biased ratios were recorded in some smaller ones. Most of these species inhabit protected sites where a few parthenogenetic species were also found.

A recorded bias in sex ratio towards females is frequently attributed to the mass death of male animals soon after mating. For example, Warburg and Cohen (1992) recorded a sex ratio approximating unity in the isopod *Armadillo officinalis* with a bias towards females. This was probably due to the death of males after mating. Similarly, adult males of the oniscid *Schizidium tiberianum* constituted less than 10% of total adults, while in juveniles the sex ratio was 1:1 (Warburg 1994). According to Warburg et al. (1993), sex ratio is a dynamic rather than constant characteristic and changes during the life history of isopods.

At first glance, the lack of amphitocous species combining both arrhenotoky (which favours rapid selection) and apomictic thelytoky (favouring rapid numerical recovery) appears surprising in ecosystems suffering from destructive human impacts such as fire. Nevertheless, in ecosystems where a modest recovery time results in progressive restoration of the initial ecosystem type, neither rapid numerical recovery nor rapid selection are advantageous. Consequently, amphitoky does not serve Mediterranean arthropods well.

5.1.3.2
Semelparity and Iteroparity

The great majority of adult Mediterranean arthropods lay their eggs early in their development. For example, the oribatid *Scheloribates* cf *latipes* with more than 1 year of adult life lays its first eggs 7 weeks after reaching adulthood. The millipede *A. t. judaicus* from mesic habitats, which has a life span of 11 years, also deposits eggs 4 months after reaching adulthood.

As to reproductive patterns, Warburg et al. (1993) report that most of the isopods from Israel are iteroparus, and only two local species, a desert and a Mediterranean one, are semelparus. Moreover, different isopod species inhabiting the same habitat differ in their reproductive patterns (Warburg 1992), and these differences are associated with anatomical differences in the ovaries (Warburg 1994). *S. tiberianum* is short lived, breeds in spring, and is semelparus, realising a considerable number of mancas before dying, whereas *A. officinalis* can survive for 9 years (or probably more), breeds in the fall, and is iteroparus. It seems that in isopods semelparity and iteroparity are equally successful reproductive strategies in the Mediterranean regions of Israel.

Iteroparity is generally considered to be the most economic reproductive strategy, since specific morphological adaptations such as physogasty or swelling of the abdomen, high energy expenditure in short periods (affecting survival), as well as extra overwintering or oversummering adaptations are needless (Siepel 1994). According to Warburg (1987a), in variable environments such as the Mediterranean ones iteroparity is favoured whenever the probability of adult survival is greater than that of juveniles.

In general, Mediterranean arthropods show traits such as concentrated mortality of immature individuals, limited body size, low metabolic rate, and a relatively long adult life span, which characterises iteroparus species. Nonetheless, it seems probable that climatic predictability may outweigh disadvantages related to temporal uncertainties, and deviation from iteroparity to circumscribed oviposition is favoured. Thus, notwithstanding the above characteristics, compromising patterns are favoured in Mediterranean regions. Most arthropods appear precocious and the main bulk of eggs is deposited during brief periods of time, while a slight distribution of reproductive effort upon different adult age groups also occurs. Deviations from iteroparity do not even increase energy costs. In fact, due to low Q_{10} values and efficient thermoregulation over the optimal temperature range, Mediterranean arthropods may significantly reduce the energy cost of reproduction.

The phenomenon of egg laying in Mediterranean arthropods is well exemplified by three oribatid and two collembolan species from Hortiatis (Figs. 5.2, 5.3). In all cases the animals deposit their eggs soon after reaching

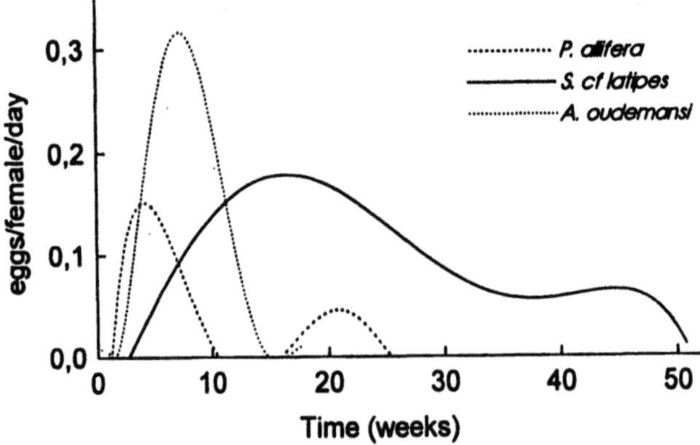

Fig. 5.2. Fecundity of oribatids *Scheloribates* cf *latipes*, *Pilogalumna allifera* and *Achipteria oudemansi* in the laboratory. Data were fitted by least squares

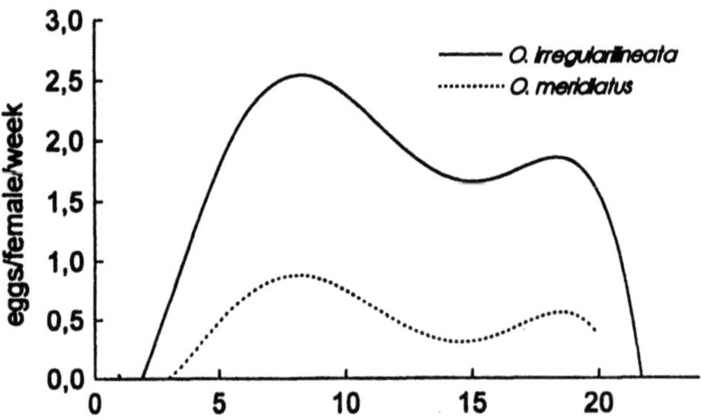

Fig. 5.3. Fecundity of the collembolans *Orchesella irregularilineata* and *Onychiurus meridiatus* in the laboratory. Data were fitted by least squares.

maturity. Although reproductive effort is to some extent dispersed over several age groups of adults, reproduction is mainly restricted to short periods and most eggs are laid by young adults aged between 5 and 25 weeks. Thus, oribatids and collembolans from Hortiatis can hardly be characterised as either semelparus or iteroparus. The millipede *Ommatoiulus moreletii* of Australian Mediterranean type regions also displays an analogous reproductive strategy. In fact, Baker, although unable to demonstrate iteroparity in laboratory populations (1978a), reported that postbreeding adult survival was inversely correlated with breeding success (Baker 1978b). Hence, although he classifies *O. moreletii* as a semelparus species, he accepts distribution of the breeding risk over time.

Mediterranean arthropods inhabit environments where seasonal predictability dominates over hazardous fluctuations of environmental variables. Nonetheless, as indicated, energy for reproduction is unevenly allocated among the different age groups. Younger and, to a lesser extent, older adults contribute more to reproduction than specimens in intermediate life stages, and relevant graphs display first and second order peaks. Dispersion of reproductive effort among the age groups can be considered to be an adaptation· to spatially heterogeneous Mediterranean habitats. Moreover, slightly distributing the reproductive effort over time can equally be viewed as an adaptation to minor climatic hazards and can be effective in response to unpredictable anthropogenic impacts. Thus, in accordance with Norton (1994), it can be concluded that Mediterranean arthropods compensate precocity and small clutch size with elements of a "bet-hedging" reproductive strategy such as long adult life, low adult mortality, and slight dispersion of reproductive effort over time, which can be considered buffers against environmental disasters.

5.1.3.3
Parental Care

Parental investment, which is linked to the survival of subsequent instars, is seen in most Mediterranean arthropods. Brood protection is common among arthropods from a variety of habitats. For example, Bercovitz and Warburg (1985) observed this phenomenon in the millipede *A. t. judaicus* and Matthiessen (1991) in the white-fringed weevil *G. leucoloma*.

Brood protection among Mediterranean arthropods involves adaptations either to predation or to drought. Relevant to the former is the fact that the eggs of oribatids and collembolans can escape the attention of hungry predators when deposited in small holes (Stamou and Asikidis 1992). Adaptations to summer drought appear to be even more significant. Apart from anatomical structures such as the marsupium in isopods, relevant adaptations can also be observed in ethology. Crawford et al. (1987) reported that maternally formed pellets surround diplopod eggs and early stages of development in both mesic and xeric Mediterranean areas. The eggs of most Mediterranean diplopods are deposited in specific chambers made either exclusively of faeces or a mixture of faecal, humic and mineral materials. Egglaying chambers presumably to some extent protect eggs from desiccation. Moreover, batches of diplopod eggs are laid in the vicinity of sites where favourable hygric conditions prevail (Baker 1978a; Bercovitz and Warburg 1988; Iatrou 1989). Similarly, most oribatids and collembolans lay their eggs on moss leaves where evapotranspiration can protect them from desiccation, while the eggs of other arthropods are laid among fallen leaves or lichen tissues.

Indirect energy allocation of another type by deceased mothers to their offspring has been described in the woodlouse *S. tiberianum* from Israeli

Table 5.2. The effect of precipitation on the recruitment to population in the diplopod
G. *balcanica.* (Iatrou and Stamou 1991)

Year	Oviposition period	Precipitation (mm)	Density of 1st stage (ind m^{-1})
1983	April-July	182.3	262.3
1984	April-July	90.4	25.69
1985	April-July	56.9	8.66
1986	April-July	189.5	381.08

Mediterranean habitats (Warburg et al.1993), which is a habitat-independent
species (Warburg 1994). In both mesic and xeric habitats, the female dies
soon after release of the mancas, the latter feeding on the female's carcass.

5.1.3.4
Reproductive Effort

Depending on habitat conditions, relatively low reproductive effort is ex-
pended by Mediterranean arthropods. Bercovitz and Warburg (1985)
recorded energy investment in eggs of the diplopod *A. t. judaiacus* to be
lower in mesic habitats than in xeric ones. In addition, Crawford et al. (1987)
and Bercovitz and Warburg (1988) recorded lower numbers of mature
oocytes in the ovaries, lower numbers of laid eggs, and smaller clutch sizes in
mesic populations of *A. t. judaicus* than in xeric ones. Among isopod populations from xeric habitats, Warburg (1994) also recorded greater numbers of
released mancas than in populations from mesic habitats. In the former,
higher reproductive allocation and higher parental investment is counterbalanced by the female's shortened life.

No relationship between reproductive allocation and reproductive strategy was revealed. Both semelparus and iteroparus species show relatively
low reproductive investment. In contrast, temperature and humidity conditions affect reproductive effort significantly. Baker (1978a, b) related reproductive effort of the Australian Mediterranean millipede *O. moreletii* to temperature and moisture conditions, and Iatrou and Stamou (1991) correlated
successful breeding of *G. balcanica* with rainfall during the oviposition period (Table 5.2).

5.1.3.5
The Effect of Temperature and Humidity on Egg-Laying Patterns

Moisture conditions are decisive for the egg-laying schedules of arthropods.
Indeed, egg deposition during the drought season requires that eggs be capable of absorbing water from moist surfaces, whereas excessive water may
reduce oxygen supply to eggs, thereby causing failures in embryogenesis
(Bercovitz and Warburg 1988). Thus, egg laying usually occurs during peri-

ods of modest moisture. Warburg et al. (1984) report that in isopods from more xeric habitats, reproduction and further oscillations in population numbers are related mostly to air humidity and rain. By contrast, in mesic habitats temperature is the factor modulating the duration of oocyte development and the onset and duration of egg laying. Moreover, unlike millipedes from more xeric habitats where moisture is the prevailing factor, temperature conditions are of greater importance for egg laying *G. balcanica* inhabiting a subhumid habitat. Under milder moisture conditions recruitment is conditioned by temperatures having prevailed in the organic layers of the soil during egg synthesis 8 months earlier. The onset as well as duration of egg deposition are strongly dependent on moderate temperatures varying from 7 to 25 °C (Iatrou and Stamou 1989b). In addition, unlike the instars of their relatives in more arid regions, egg hatching and emergence of the first free instars from the egg capsules or marsupium of Mediterranean arthropods are temperature rather than moisture dependent (e.g. Crawford et al. 1987). In any case, egg synthesis, oviposition and hatching in Mediterranean regions occur during brief periods, thereby contributing to synchronisation of life cycles with the seasonally or interannually varying environment. (Fig. 5.4).

5.2
Synchronisation Tactics

Environmental thresholds, in conjunction with age at maturity and voltinism are of crucial importance for the synchronisation of arthropod activity (recruitment to population, interruptions in life cycle development etc.) with the fluctuating environment. Univoltine species with obligatory diapause and/or dormancy, and bivoltine species with facultative aestivation and dormancy ensure synchronisation of the life cycle of epiphytic arthropods with seasonally varying climatic variables.

Voltinism is also important for synchronisation in soil-dwelling arthropods. According to Siepel (1994), it can be stated that high reproductive success and circumscribed oviposition accompanied by winter quiescence and/or aestivation are the principal correlates of seasonal rhythms in Mediterranean regions. Circumscribed oviposition and facultative aestivation or quiescence enable animals to easily overcome unfavourable periods of the seasonal cycle.

Most soil inhabitants carry their eggs for long periods and can lay them opportunistically whenever environmental conditions are temporarily within favourable limits. Thus, circumscribed field oviposition, which is beneficial in seasonally varying environments, is the by-product of environmentally conditioned schedules of oviposition. Moreover, true diapause or dormancy involving morphological and physiological changes are uncommon in soil arthropods, and only a few collembolan species enter ecomorphosis or any-

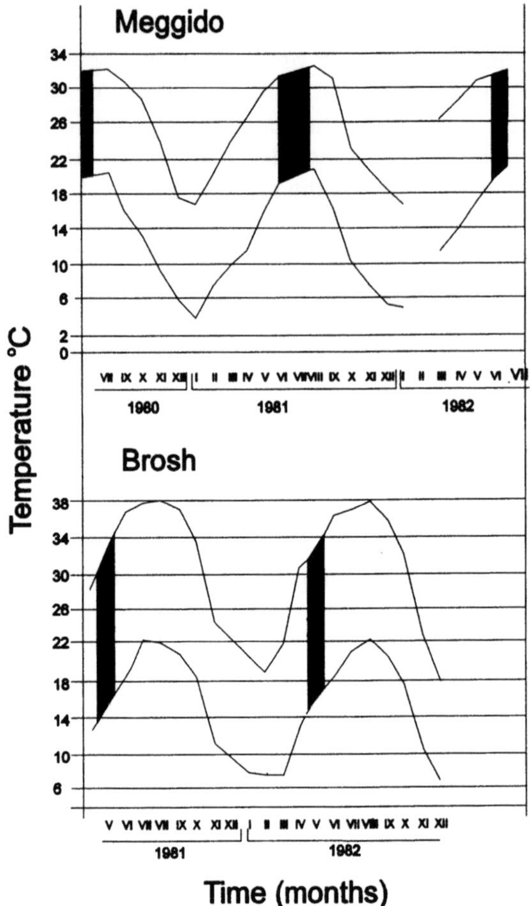

Fig. 5.4a,b. Monthly minimum and maximum temperatures of xeric and mesic habitats in Israel. *Shaded area* indicates duration of egg laying period of the diplopod *Archispirostreptus tumuliporous judaicus.* (Bercovitz and Warburg 1988)

drobiosis. Most soil inhabitants have continuous generations, and aestivation occurs in the egg stage or during moulting, while suppressed metabolism leading to quiescence is imposed below certain threshold temperatures in winter.

It seems plausible that being able to conform to predictable seasonal cycles may be more important for species dominance than per se values of demographic parameters. In most cases, the rate of population increase is not directly linked to species dominance in the field. As a mater of fact, Iatrou and Stamou (1991) combined field and laboratory data and compiled a life table for the dominating millipede on Hortiatis, *G. balcanica* (Table 5.3). Far higher mortality is exhibited among immature individuals than among those in reproductive stages in which mortality is low. The overall result is that the

Table 5.3. Life table for the diplopod *G. balcanica* dominating in the organic layers at the evergreen-sclerophyllous formation at Hortiatis. (Iatrou and Stamou 1991)

Age structure	n_x	l_x	m_x	$l_x m_x$	k Values
Egg	382.00	1.00			
1st stage	90.00	0.236			0.63
2nd stage	73.16	0.192			0.36
3rd stage	39.10	0.102			0.27
4th stage	18.11	0.047			0.34
5th-8th stage	4.66	0.012			0.00
9th-11th stage	4.66	0.012			0.00
12th-13th stage	4.66	0.012	9.00	0.108	0.00
14th stage	4.66	0.012	13.80	0.166	0.00
15th stage	4.66	0.012	22.80	0.0.274	0.00
16th stage	4.66	0.012	16.20	0.194	0.00
17th stage	2.67	0.007	19.80	0.139	0.23
18th stage	1.06	0.003	16.20	0.049	0.37
19th stage	0.19	0.0005	12.00	0.006	0.78
				$R_o=0.94$	

n_x, Number of females; l_x, probability of survival; m_x, fecundity

net rate of reproduction results in a value approximating 1 (R0=0.94), which hardly ensures stable age distribution.

In summing up the above discussion, it can be stated that in comparatively less severe, although strongly fluctuating Mediterranean environments, periods favourable to energy consuming activities, such as the synthesis and deposition of eggs and their subsequent development, are short. Hence, relatively few eggs moderately endowed with yolk are rapidly synthesised and deposited, whereas accumulation of immature specimens (population reserves) can be recorded in winter due to low heat budgets. Furthermore, although predictable climatic components dominate over hazardous ones, animals are capable of overcoming environmental hazards by slight dispersion of oviposition risks over time.

5.3
Life Cycle Development

5.3.1
Life Cycle Development of Short-Lived Arthropods

The above conclusions can be exemplified by oribatids from Hortiatis. Figure 5.5, which has been redrawn from original graphs provided by Asikidis and Stamou (1992), shows the population dynamics of the oribatid *Scheloribates* cf. *latipes*. A population dynamics model (Stamou 1986a) was fitted on

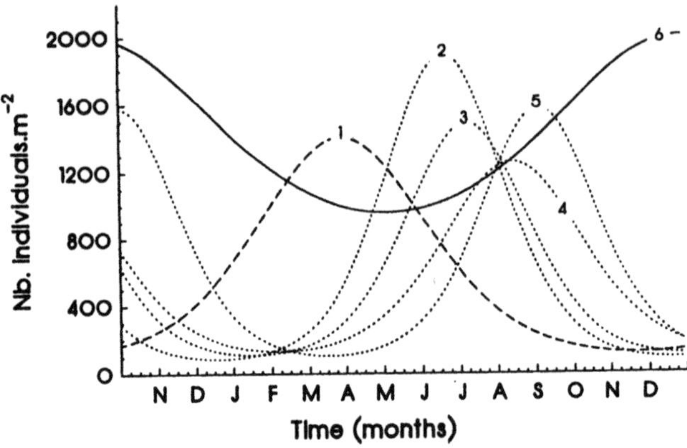

Fig. 5.5. Population dynamics of the oribatid *S. cf latipes*. The model proposed by Stamou (1986a) is fitted on census data. For the sake of simplicity only theoretical values are depicted)

the original data. The conclusions drawn from evaluating the model's parameters (Stamou and Sgardelis 1989) are as follows: although a preponderance of mature individuals was recorded throughout the year, adults are most numerous in late autumn and early winter, and the peak in density occurs in November/December. As in most oribatids from temperate regions (Luxton 1981b), the adult is the overwintering stage on Hortiatis. *S. cf. latipes* is univoltine, initiating one generation per year. Eggs develop and are laid soon after the animals attain maturity. The greatest numbers of eggs are to be found in the abdomens of females in March, while the oviposition period lasts from February to June. New larvae first emerge in April, and during summer the immature forms develop rapidly into subsequent life stages. As shown in the model, adults appear to be precocious, and reproductive effort is distributed over a short period of time. These two features are typical of populations inhabiting seasonal environments. Furthermore, the development of juvenile instars follows the seasonality of the environment. They develop during short specific periods, while the overall life cycle shows annual periodicity.

Pilogalumna allifera and *Achipteria oudemansi*, the other characteristic species, share the above characteristics of *S. cf. latipes*. However, their life cycles follow a different timing. The eggs withstand summer drought by remaining dormant and hatch in early autumn. Juveniles develop rapidly in the autumn and adults emerge in early winter. Moreover, *P. allifera* lays eggs in open sites and appears seasonally iteroparus, ovipositing twice a year, i.e. in early spring and in autumn. Autumnal eggs overwinter and develop along with those laid in the following spring. Thus, seasonal iteroparity spreads the risk of abiotically unpredictable disasters in exposed sites, since even the complete loss of a reproductive season is allowed.

Notwithstanding similarities in life history characteristics such as precocity and circumscribed oviposition, the timing of the life cycles of the three oribatid species appears species specific and differs in the timing of different activities. *S.* cf. *latipes,* which inhabits sheltered microsites underneath *Quercus coccifera* shrubs, can develop during summer. The eggs of *A. oudemansi,* which is restricted to semi-sheltered microsites close to the shrubs but not underneath them, withstand summer drought in a state of dormancy and develop rapidly in autumn. Finally, the eggs of *P. allifera* are laid in open sites in autumn and in spring. After emergence in late spring, the larvae overcome summer drought by remaining dormant and develop rapidly in the autumn.

Generalising the above considerations, it can be concluded that the life cycle and development of microarthropods in Mediterranean regions is under the control of strongly and largely predictable variations in climate. In specific cases, other external factors including edaphic ones also become important (Asikidis and Stamou 1992). Intrinsic factors, such as inter- and intraspecific competition, are in all probability of lesser importance in the life cycle and development of microarthropods. Analogous conclusions have been reported for animals living in other severe environments such as the Antarctic (e.g. Block 1985).

5.3.2
Life Cycle Development of Long-Lived Arthropods

The timing of the life cycle and the development of long-lived arthropods also appear to be species specific. For example, in the Mediterranean regions of Israel, the isopod *Philoscia muscorum* peaks in winter, *Metoponorthus pruinosus* in early spring, and *A. officinalis* in late spring (Warburg 1987a). However, like those of their short-lived relatives, the life cycle and development of long-lived arthropods in Mediterranean regions are synchronised with both seasonal and interannual cycles. Figure 5.6 shows a diagram of the life cycle and development of the diplopod *A. t. judaicus* in mesic and xeric habitats of Israel. Egg laying and development of the most vulnerable younger stadia occur during short periods. After 1 year the animals are at stages IV and V in the mesic and xeric environments, respectively. Animals in the subsequent stadia moult once a year in the mesic habitat and three times a year in the xeric habitat.

Obviously, the life cycle development of macroarthropods is realised differently under different habitat conditions. However, the sequential occurrence of different phenophases within short time periods – ensuring synchronisation with interannual cycles – is invariably recorded. The life cycle of the millipede *G. balcanica* on Hortiatis is illustrative. Within a period of 3.5 years, only one successful breeding was recorded in April/July 1983. The animals normally attain maturity after 3-3.5 years, and a second generation was initiated 3 years later in April/July 1986. As could be shown, unsuccessful

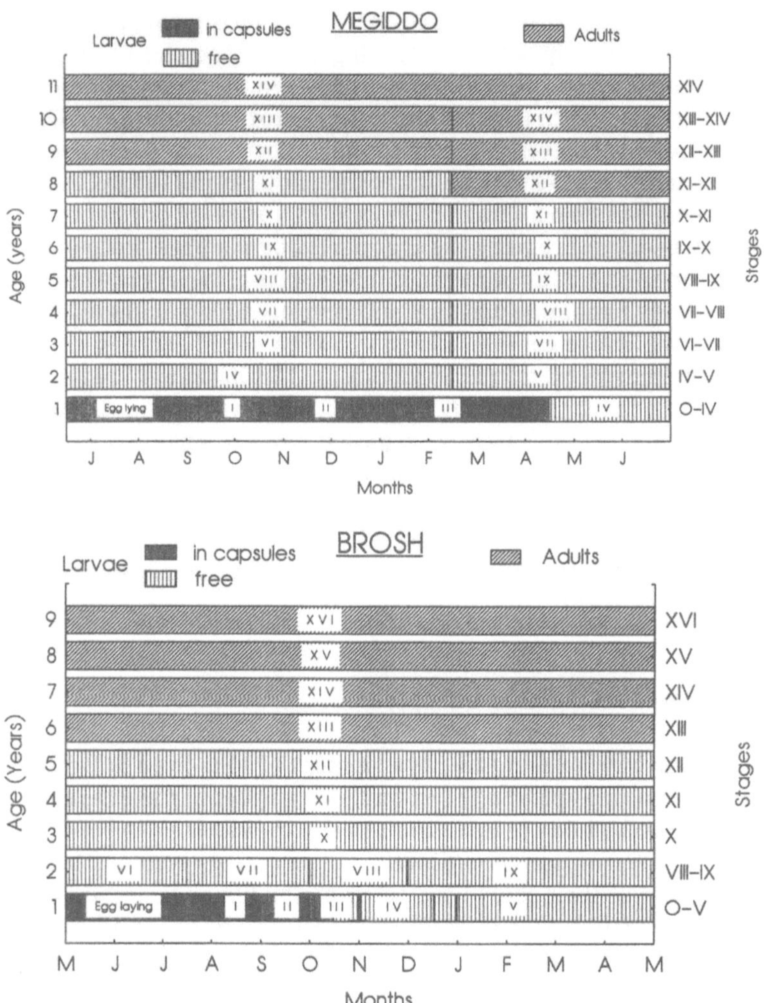

Fig. 5.6a,b. Life cycle development of the diplopod *A.t. judaicus* in a mesic (Megiddo) and b xeric (Brosh) Israeli habitats. (Bercovitz and Warburg 1985)

breeding was directly linked to low rainfall during the egg laying period from April to July. Litter consumption and production of faeces, from which egg capsules are made, depend closely on litter water content (Bertrand et al. 1987). In any case, distribution over the time of reproductive effort coupled with a clutch size depending on climatic variables results in a life cycle complying with the predictable interannual periodicity of the temperature-humidity complex.

The open question is whether – as is the case in arthropods of deciduous forests (e.g. Luxton 1981a,b; McQueen and Steel 1980) – external signals operate as indicators stimulating the timing of development. Existing evidence is controversial. Photoperiod in conjunction with temperature seems to trigger oogenesis in the oniscids of xeric Mediterranean formations (Warburg 1994). Baker (1979a) also reported that the activity of the diplopod *O. moreletii* in southern Australia is stimulated by rainfall and increasing temperature. In contrast, Bercovitz and Warburg (1988) studied the reproductive patterns of the diplopod *A. t. judaicus* in Israel and reported that egg laying occurs at temperatures between 16 and 32 °C, but the onset of egg deposition does not take place on a definite date. Moreover, they did not find any effect of the length of daylight on the onset of egg laying. Asikidis and Stamou (1992) reasoned that external triggering might be associated with the uncertainty emerging in cases where outstanding environmental hazards mask the functioning of predictable variables. In Mediterranean regions, the preponderance of the predictable components of the climate over accidental hazards is well documented (e.g. Sgardelis 1988). Hence, signals that trigger the timing of life cycle development are needless in Mediterranean environments. Instead, the existence of critical climatic thresholds – either allowing biological activity or not – may force development to occur during short periods, thereby ensuring synchronisation with seasonally varying environments.

Phenological Patterns

The temporal distribution of arthropod populations is the outcome of a complex process involving both demographic and physiological adaptations (Stamou et al. 1993a). The study of field population dynamics may therefore reveal the way in which evolutionary lineage can be realised under specific conditions, and further hypotheses regarding adaptation can be tested. The opposite is also possible: hypotheses regarding arthropod life history tactics and physiological adaptations can be formulated from census data and phenograms. This approach has been deliberately adopted for most Mediterranean arthropods, so explanations of the synchronisation of their development with seasonally varying climatic variables can be inferred.

6.1
Modelling Phenological Patterns

In most cases the phenograms of Mediterranean arthropods display asymmetries and discontinuities. For example, the phenogram of the isopod *Meta-*

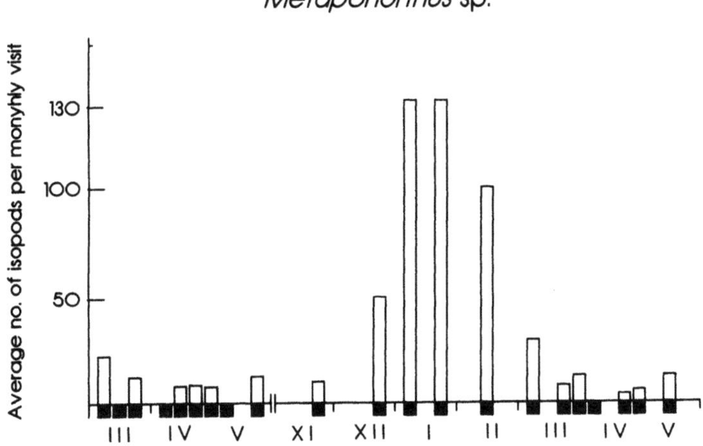

Fig. 6.1. Phenogram of the isopod *Metaponorthus* sp. from northern Israel (Warburg et al. 1984)

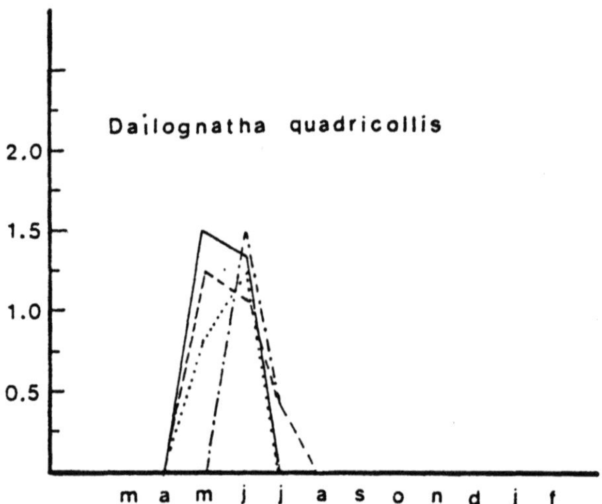

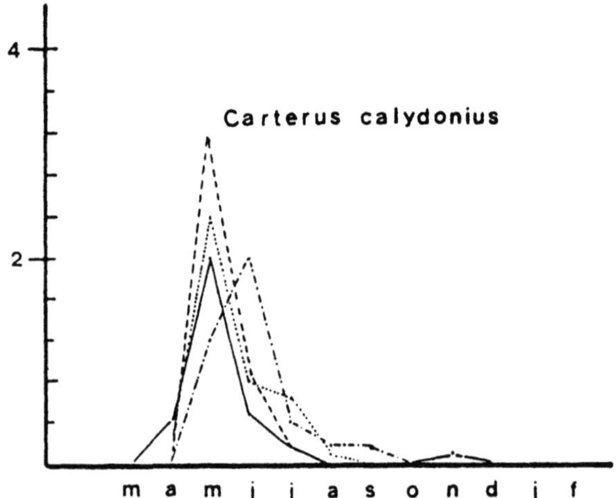

Fig. 6.2. Phenogram of two coleopteran species from four different litter types of an insular ecosystem. *X-axis* depicts months and *y-axis* the number of individusls per trap and month. *Continuous line* Open ground; *dashed line Olea* litter; *dotted line Juniperus* litter; *mixed dashed and dotted line Pistacia* litter. (Trihas and Legakis 1991)

ponorthus sp. from northern Israel appears to be skewed to the right, depicting a steep elevation in density in December/January. Afterwards, density declines gradually until May (Fig. 6.1). An analogue is the phenology of the Coleoptera *Dailognatha quadriocollis* and *Carterus calydonius* in the four different litter types of an insular ecosystem (Fig. 6.2). Right- or left-skewed or even flat phenology curves are also obtained from arthropods from an

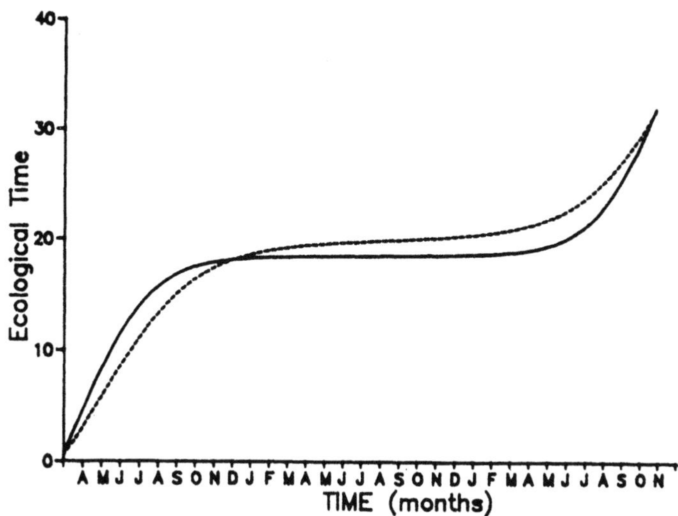

Fig. 6.3. Relationship of ecological to standard time. *Solid line* immature; *dashed line* pseudo-mature of *Glomeris balcanica.* (Stamou et al. 1993b)

evergreen-sclerophyllous formation of southern France (Bigot and Bodot 1973).

To model asymmetric phenologies, Stamou et al. (1993a) defined the unit of ecological time as the standard-clock time interval during which a constant number of demographic events (births plus deaths) occur. During population explosions, a given number of demographic events (for example ten births plus deaths) will occur in shorter standard time intervals than during other times. Thus, when defined demographically, the unit of ecological time changes during the passage of standard clock time (Fig. 6.3). Increasing parts of the curve correspond to rapid development in numbers, whereas the plateau indicates negligible changes in population size. Stamou et al. (1993a) suggested that changes in ecological time parallel changes in the environmental gradients overriding population dynamics in the field. The relationship between the two time units depends upon the demography as well as on the physiological response of each species to environmental component variables. Hence, dynamic relationships specific to each population can be formulated.

6.2
Numerical Responses of Microarthropods

A curve was fitted numerically to census data referring to the phenologies of 11 collembolan and eight oribatid species from Hortiatis (Figs. 6.4, 6.5). As

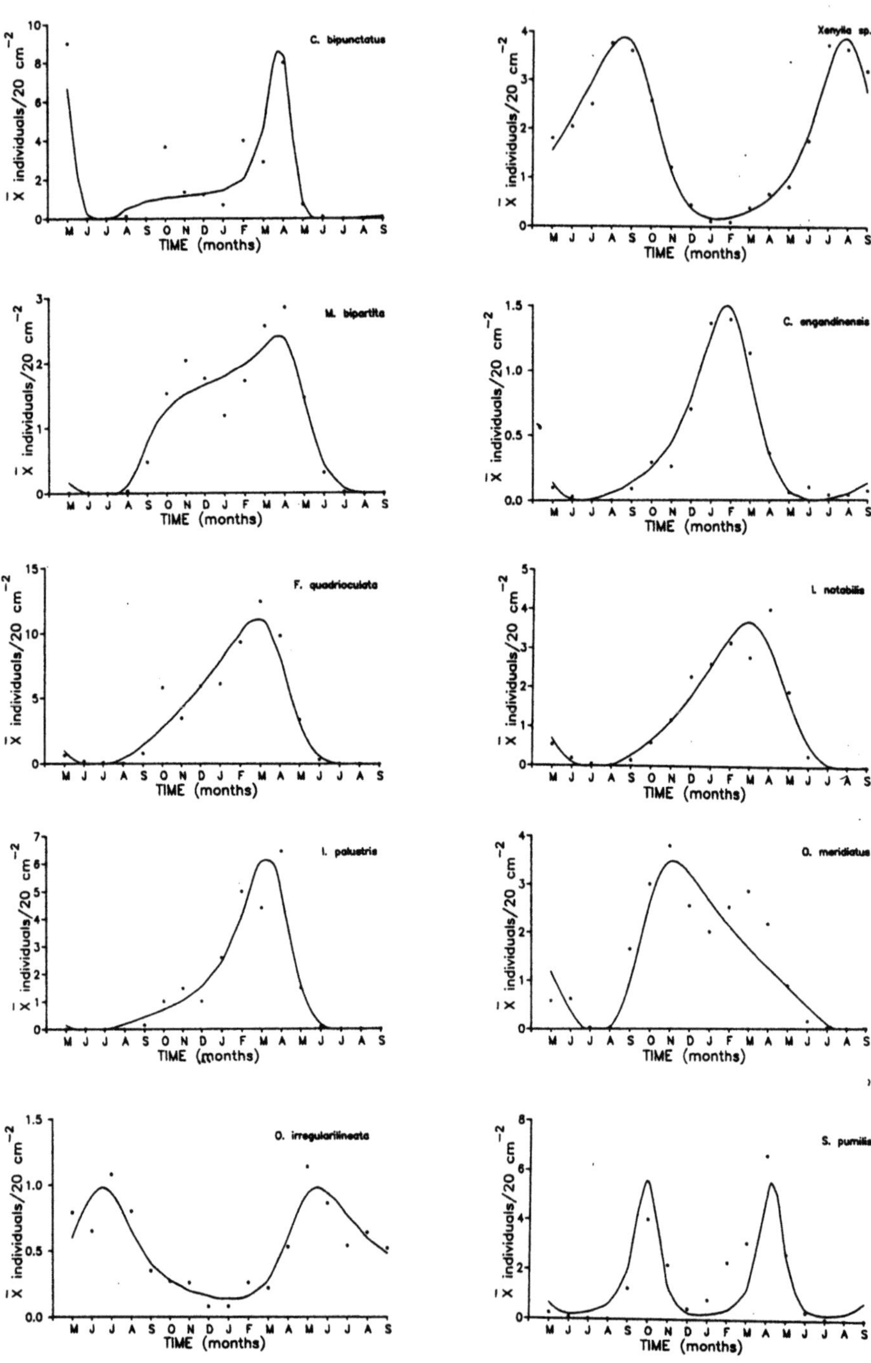

Fig. 6.4. Fitting of the phenological model to census data for collembolans (Stamou et al. 1993a)

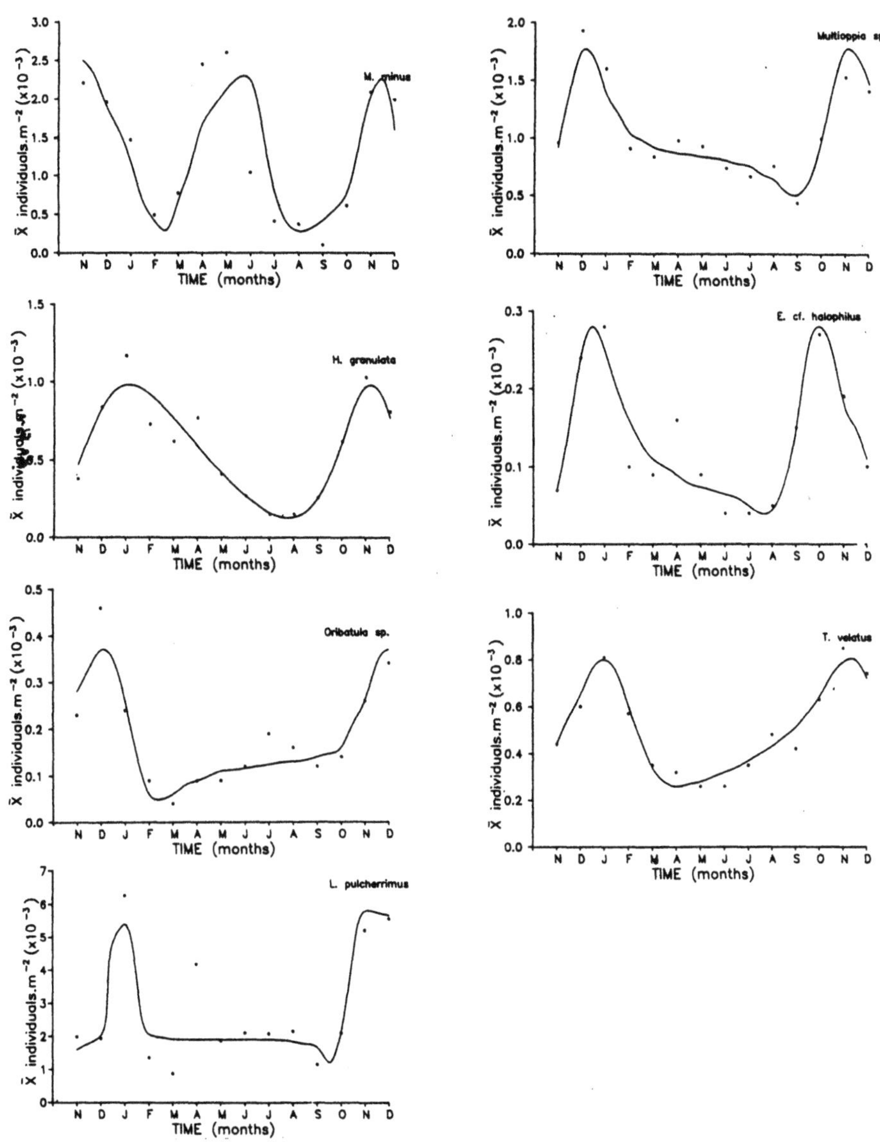

Fig. 6.5. Fitting of the phenological model to census data for oribatids (Stamou et al. 1993a)

largely predicted, seasonal fluctuations in climatic variables result in the interannual stability of the phenograms of both collembolan and oribatid populations. Moreover, the great majority of oribatids and collembolans display an annual periodicity in their life cycle, and summer is the adverse period for most species.

Table 6.1. Life history traits of collembolans and oribatids displaying left- and right-skewed asymmetric as well as symmetric phenologies. (Stamou et al. 1993a)

Left-skewed phenology	Right-skewed phenology	Symmetrical phenology
Precocity	Precocity	Precocity
Iteroparity (circumscribed oviposition)	Iteroparity (circumscribed oviposition)	Iteroparity (either circumscribed or uncircumscribed oviposition)
High rate of juvenile developmet into subsequent stadia during the favourable period	High rate of juvenile developmet into subsequent stadia during the favourable period	High rate of juvenile developmet into subsequent stadia during the favourable period
Developmet of juveniles into subsequent stadia during the unfavourable period	No developmet of juveniles into subsequent stadia during the unfavourable period	Development of juveniles into subsequent stadia during the unfavourable period
Small number of eggs overcoming adversity	Large number of eggs overcoming adversity	Either small or large number of eggs overcoming adversity
More evenly distributed reproductive effort during adulthood	Less evenly distributed reproductive effort during adulthood	Reproductive effort evenly distributed during adulthood or confined to short periods
Left-convex adult survivorship curve	Right-convex or linear adult survivorship curve	Right-convex or S-type adult survivorship curve

According to Stamou et al. (1993a), on account of the capacity of most immature oribatids to develop even during adverse periods, the cohort that overcomes a time of drought is age structured. By contrast, collembolan cohorts overcoming adversity consist either of aestivating eggs which were deposited at the beginning of drought or of adults in anhydrobiosis. As a result, the number of oribatids can recover rapidly after summer drought, and right-skewed phenologies are displayed. In contrast, collembolans exhibit left-skewed phenologies and density peaks occur later in the year. In oribatids, the energy-consuming capacity to withstand summer drought is distributed over the various the life stages and is compensated for by relatively low mortality rates. The reverse holds for Collembola.

The life history tactics of species displaying skewed phenologies are shown in Table 6.1. The capacity of most oribatid populations to recover rapidly during autumn results from the hatching of eggs which were deposited in large numbers before the beginning of the adverse period and from the rapid development of immature individuals which attain adulthood within short periods of time. In turn, the newly emerged adults produce and deposit eggs rapidly. The smooth decline in population size corresponds with a lower reproductive value among subsequent adult stages which is coupled with a left-convex or linear survivorship curve for these adults.

Condensed ecological time coincides with decreasing population size in most collembolan species, and phenologies are right-skewed. These species are adapted to respond rapidly to adversity. The smooth increase in popula-

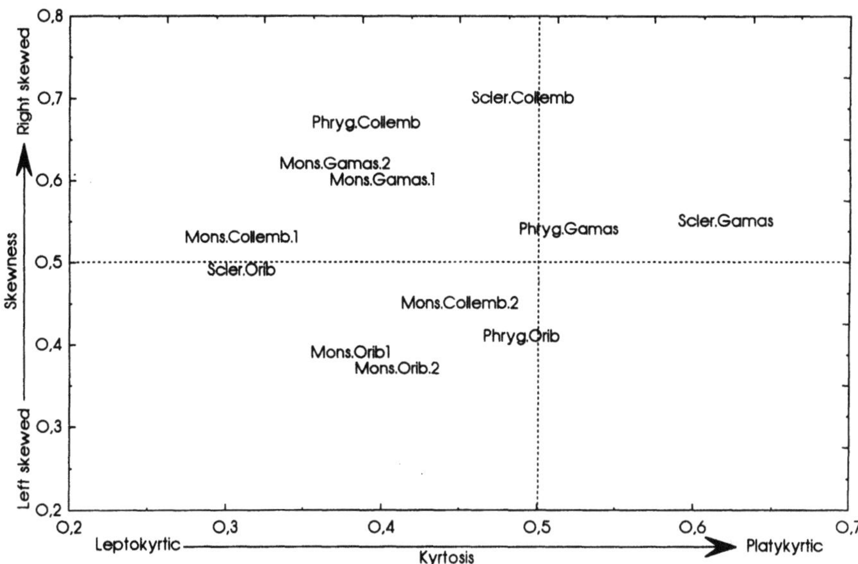

Fig. 6.6. Ordination of phenograms on the skewness-kurtosis plane. *Phryg* Phryganic ecosystem; *Mons* monsoon ecosystem; *Scler* evergreensclerophyllous ecosystem; *Collemb* collembolans; *Orib* Oribatids; *Gamas* Gamasidae. (Redrawn from Sgardelis et al. 1993)

tion size indicates a small number of eggs overcoming adversity, coupled with a high rate of juvenile development and to some extent with a broadly distributed reproductive effort throughout adulthood. A right-convex survivorship curve accounts for the rapid decline in population size during the transition from a favourable to an adverse period.

Few species exhibit either acute or flat but symmetric phenologies. Those with such phenologies are characterised by the rapid development of juveniles and right-convex or S-type adult survivorship curves. The reproductive effort can either be confined to short periods giving rise to sharp phenograms or be evenly distributed over the adult classes, thus generating flat phenograms. The number of eggs capable of overcoming adversity can also be low or high.

Precocity, circumscribed oviposition, and the rapid development of juveniles during favourable periods are features shared by all three phenological types. The characteristics distinguishing phenological patterns from one another relate to the temporal distribution of reproductive effort, adult survivorship, and the timing of oviposition.

To acquire more precise information, Sgardelis et al. (1993) combined model parameters and estimated the statistical kurtosis and skewness of the phenograms. For comparative purposes, the method was used to explore phenologies from a monsoon, evergreen-sclerophyllous and phryganic ecosystem (Fig. 6.6). Three phenological types were distinguished: left-skewed leptokyrtic (oribatids), right-skewed platykyrtic (Mediterranean Gamasidae) and right-skewed leptokyrtic (Mediterranean Collembola and Gamasidae from India). In both Mediterranean and monsoon ecosystems, adversity engenders asymmetric phenologies. However, the phenological responses reflecting avoidance-tolerance strategies are taxon-specific as well as region-specific. Oribatids respond to adversity in the same way in both the Mediterranean and monsoon ecosystems. Numbers are relatively low at the beginning of an adverse period, while immature instars develop rapidly during and/or immediately after the adverse period. Gamasidae display right-skewed phenograms in both regions, although differences in kurtosis are revealed. In Mediterranean ecosystems, Gamasidae seem to tolerate adversity and their relevant phenograms are platykyrtic, whereas in India they are leptokyrtic. Collembolans also display region-specific phenograms. As mentioned above, Mediterranean Collembola display left-skewed phenologies, while the corresponding phenologies in India are right-skewed.

Skewed phenologies apparently characterise arthropods encountering adverse periods. For example, unlike their Mediterranean and monsoon counterparts, Collembola from a temperate forest (Fontainebleau, France; Sgardelis et al. 1993) respond to unfavourable seasons by migrating through both the organic and mineral layers of the soil. In winter, they are abundant in the mineral layers and migrate upwards in spring.

6.3
Numerical Responses of Macroarthropods

The phenologies of long-lived arthropods must be synchronised with both seasonally varying climatic variables and interannual cycles. As shown above, in long-lived diplopods such as *G. balcanica* from Greece, *A. t. Judaicus* from Israel and *O. moreletii* from Australia, synchronisation with interannual cycles is ensured by the timing of oviposition. In contrast, the mass development of short-lived immature stadia is related to seasonality and confined to more or less short periods of time. Finally, long-lived adults display seasonal activity patterns. Figures 6.7, 6.8 and 6.9 show phenograms of the immature stadia I, II, III and IV and of the pseudomature stage and adults of *G. balcanica*. Immature stages appear to be susceptible both to the seasonality of the Mediterranean climate and to the 3- to 4-year interannual climatic cycle. Mass development of immature individuals into subsequent stadia occurs over short periods. Although constantly present in samples, they are recorded in low numbers due to unsuccessful recruitment to the population, while peaks in their density follow the 3-year interannual cycle. Adult numbers display a clear-cut left-skewed seasonal pattern, and dense ecological time coincides with the moulting period in summer. Adults thus respond rapidly to

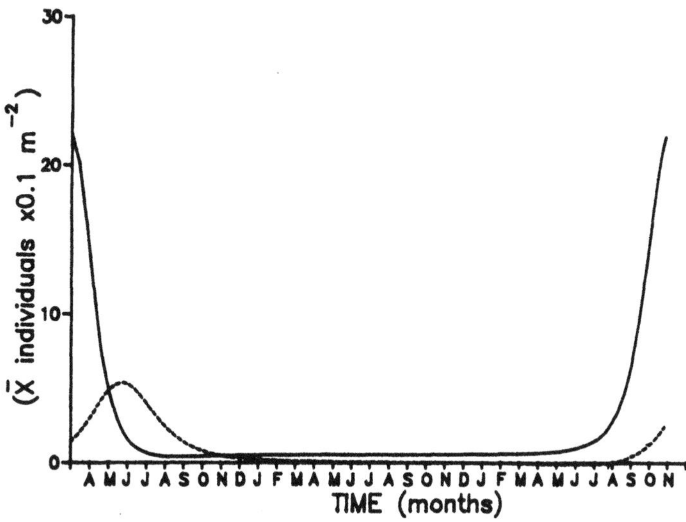

Fig. 6.7. Fitting of the phenological model to census data for *Glomeris balcanica* stage I (*solid line*) and stage II (*dashed line*) (Stamou et al. 1993b)

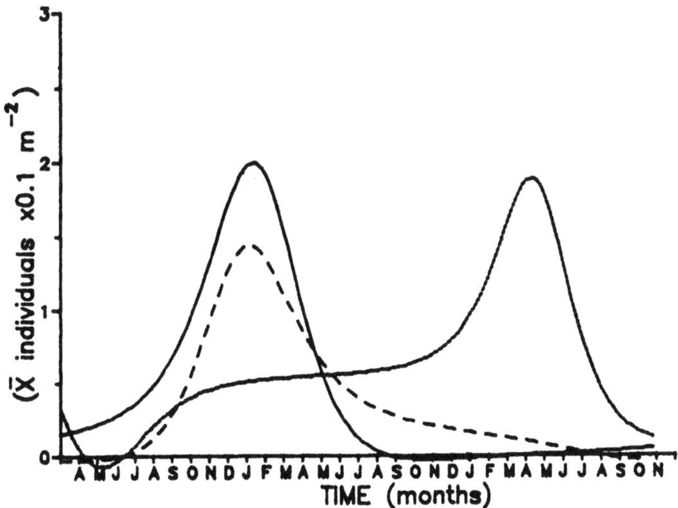

Fig. 6.8. Fitting of the phenological model to census data for *G. balcanica* stage III (*solid line*), stage IV (*long dash*), and pseudomature (*short dash*) individuals. (Stamou et al. 1993b)

summer drought by entering moulting. Moreover, unlike immature stages, adults appear to some extent to be susceptible to winter stress. The peak in density occurs in early summer.

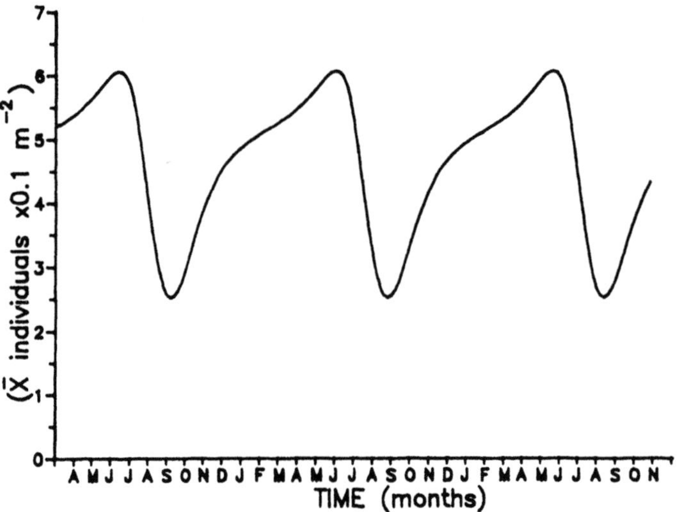

Fig. 6.9. Fitting of the phenological model to census data for adult *G. balcanica*. (Stamou et al. 1993b)

6.4
Numerical Responses to Environmental Disasters

6.4.1
Population Dynamics

Apart from modest human interventions, such as controlled burning or grazing, Mediterranean ecosystems have recently suffered from man-induced disasters which have had catastrophic effects on arthropod numbers. Uncontrolled fires in Mediterranean pine forests provide a good example. During the past few decades, exploitation of pine woods for tree sap, timber and combustible materials has decreased dramatically. In consequence, dense tree canopies and shrubby understory ones have developed and flammable materials have accumulated on the ground surface. Consequently, wild fires cause almost complete destruction of huge areas, while the ground surface reaches high temperatures. It is uncertain as to what extent arthropods, which are adapted to overcoming modest environmental stress, manage to withstand such severe environmental disasters.

To explore the abilities of Mediterranean arthropods to recover from modest as well as disastrous disturbances given demographic parameters and life history features, Stamou and Stamou (1996) simulated population dynamics using a Leslie-type population dynamics model (e.g. Cancela da Fonseca and Hadjibiros 1977; Stamou 1981). This task was undertaken in the light of the ecophysiological and population features of the species concerned (e.g. heat budgets, length of life cycle, etc.) which are the principal determinants of their ability to withstand stress.

The Leslie-type model accounts more for biological realism than for precision. It is therefore a suitable tool for exploring unknown situations given life history traits and physiological characteristics as they are reflected by demographic parameters. Model implementation presupposes the estimation of the demographic parameters fecundity f, mortality m, and duration of ontogenetic development l under different conditions. It further presupposes estimations of the distribution of reproductive effort and of the onset of egg production. Thus, it provides the information necessary for further analysis of life history tactics.

6.4.2
Modelling Population Dynamics

The structure of the Leslie-type model is as follows:

$$N_{t+1} = AN_t$$

where N_t and N_{t+1} are demographic vectors describing the age structure of the population at moments t and $t+1$ and A is the transition matrix. The structure of the demographic vectors and the form of the transition matrix are as follows:

$$
N_t = \begin{bmatrix} N_1 \\ N_2 \\ .... \\ .... \\ N_i \end{bmatrix}
\qquad
A = \begin{bmatrix}
0 & 0 & .. & f_i \\
S_2 & 0 & .. & 0 \\
P_2 & S_3 & .. & 0 \\
.. & .. & .. & .. \\
0 & 0 & P_{i-1} & S_i
\end{bmatrix}
$$

where N_1, N_2, N_i are the densities of the stadia 1,2,i and f_i is fecundity. The elements S_i and P_i of the transition matrix are estimated as follows from the demographic parameters mortality m_i and duration l_i of the life stage $_i$:

$$S_i = (1-m_i)\frac{l_i-1}{l_i}, \quad P_i = \frac{1-m_i}{l_i}$$

To account for the seasonality of the Mediterranean climate, all three demographic parameters are considered to be temperature dependent (Chap. 3), while only mortality and fecundity were considered moisture dependent.
The principal constraint on the implementation of the above model is its large number of parameters (12 for each life stage). To overcome this difficulty, Stamou and Stamou (1996) adopted a "fuzzy" system approach. Passing over details, it is obvious that fuzzy modelling allows for a realistic and general, though only proximate solution of over-parametrised situations by

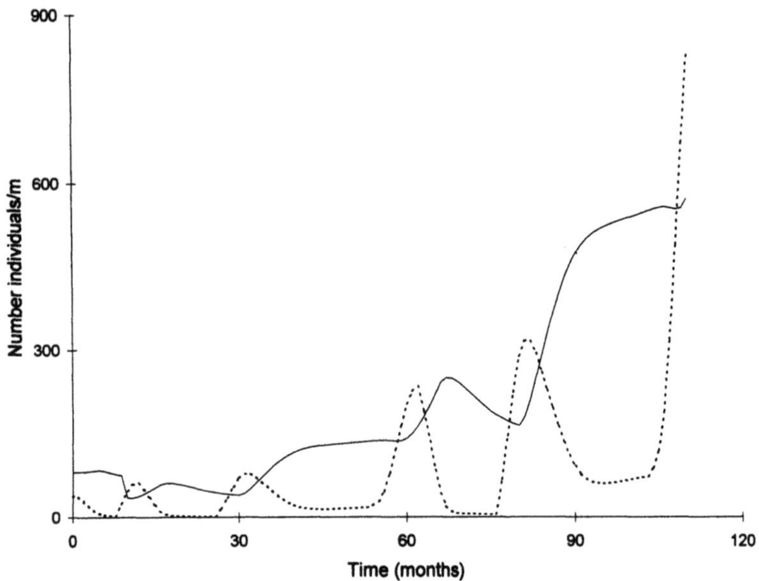

Fig. 6.10. Simulation of the population dynamics of the oribatid *Scheloribates* cf. *latipes* under regular seasonal fluctuations of temperature and humidity. *Solid line* represents adult and *dashed line* immature stages.

reducing the complexity of the system. The great advantage of fuzzy modelling lies in the fact that instead of arithmetic variables, it entails the use of linguistic terms such as "very low temperature", "low", "medium", "high moisture" etc. (Terano et al. 1992). In this way, the complexity of the overmathematised system is reduced. The highly difficult task of defining the dynamics of the demographic parameters is avoided, because they are described qualitatively by using linguistic variables and rules. Moreover, this approach generates more realistic descriptions of the ecological situation. For example, fuzzy modelling makes it possible to delineate situations in which an average recorded temperature of 20 °C is actually 20 °C in some microsites and 15 or 25 °C in others.

Different thermal and stress situations have also been simulated. They show: precocious species which to some extent distribute spread reproductive effort over time and exhibit a modest reproductive potential, rapid development of immature stages, relatively long life spans, and relative independence of fluctuating temperatures can withstand relatively heavy environmental pressure. Populations showing such characteristics (i.e. the majority of Mediterranean oribatids and collembolans for which data are available) are unstable under smooth seasonal oscillations, and numbers explode every 2 years (Fig. 6.10). In contrast, large hazardous fluctuations in spring and autumn superimposed on the seasonal pattern result in interannual popula-

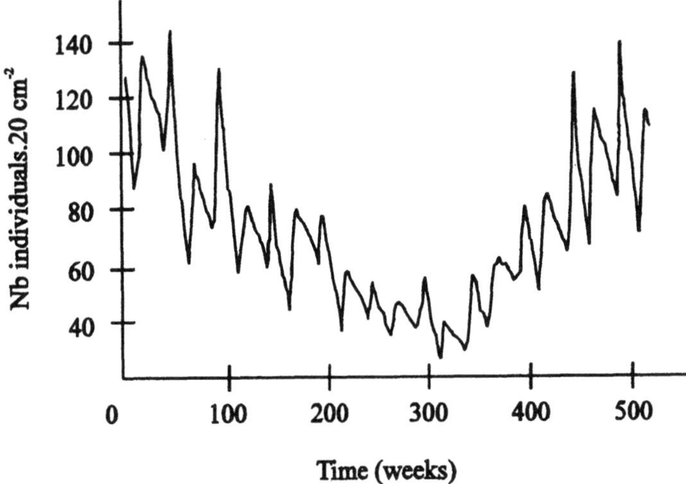

Fig. 6.11. Simulation of the population dynamics of the oribatid *S.* cf *latipes*. Violent fluctuations of temperature (± 10 °C) in spring and autumn are superimposed on regular seasonal fluctuations of temperature and humidity

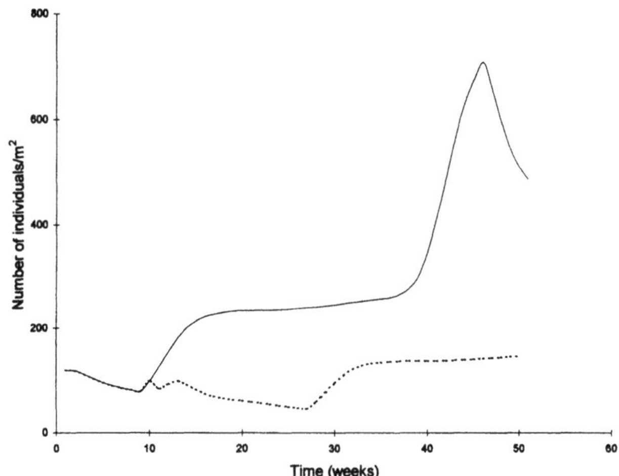

Fig. 6.12. Simulation of the population dynamics of the oribatid *S.* cf *latipes* under regular seasonal fluctuations of temperature and humidity (*solid line*). The recovery time after an environmental disaster occurring in autumn and causing an immediate 90% reduction in *numbers* (*dashed line*) is about 2 years

tion cycles (Fig. 6.11). Given the physiological and demographic characteristics, it appears that violent climatic fluctuations in autumn and/or spring are a prerequisite to interannually stable population dynamics in Mediterranean regions. The recovery time of arthropod populations after environmental disasters occurring during the favourable season and causing an immediate 50% reduction in numbers is about 1 year. Recovery after disasters causing 90% reduction in numbers takes about 2 years (Fig. 6.12).

CHAPTER 7

Community Structure

7.1
The Composition of Arthropod Communities

Despite their heterogeneous landscapes which are associated with strongly fluctuating climatic variables, Mediterranean ecosystems support more or less homogeneous faunas rich in both numbers of species and individuals. According to Vitali-Di Castri (1973), the faunas in Mediterranean regions were established prior to the development of the Mediterranean climate. Most arthropod species inhabiting regions with Mediterranean climates are therefore cosmopolitan (Table 7.1), while endemic species are somewhat rare (e.g. Sgardelis et al. 1981; Greenslade 1982; Bernini 1984; Poinsot-Balaguer 1988). In general, specific Mediterranean faunas such as those of dung beetles over the Iberian peninsula (Kirk and Ridsdill-Smith 1986) have rarely been recorded. Moreover, different Mediterranean habitats have species in common, and habitat-specific faunas have rarely been reported. For example, Warburg et al. (1980) recorded overlaps in the distribution of scorpion species over mesic and xeric Mediterranean regions as well as in the arid regions of Israel.

Table 7.1. Comparison of the faunal characteristics of arthropods from temperate, Mediterranean and desert regions (Modified from Poinsot-Balaguer 1978)

Faunal characteristic	Temperate regions	Mediterranean regions	Desert regions
Origin	Cosmopolitan	Cosmopolitan	Cosmopolitan
Species richness	High	High	Low
Abundance	High	High	Low
Faunal composition	Balanced	High, predominance of oribatids	Ddominance of prostigmatids
Seasonality	Low amplitude	Strong oscillations, phenological peaks during the rainy season	Important fluctuations, phenological peaks after rainfalls
Species diversity	High	High, strongly oscillating	Low, peaks after the rainy season

Di Castri (1973) as well as Di Castri and Vitali-Di Castri (1981) reported values of species diversity intermediate between those of rain forests on the one hand and desert or antarctic habitats on the other. Di Castri (1973) also stated that the average density of arthropods in Mediterranean-type ecosystems is intermediate between the values of moderate and severe environments. Field data seem to corroborate the above suggestions. For example, Postle (1985) reported arthropod densities in the range of 13 700-39 400 and 1300-34 000 individuals m^{-2} in the soil and litter of an Australian jarrah forest, respectively, while Sgardelis and Margaris (1993) recorded about 49 000 surface and soil arthropods m^{-2} in a Greek phrygana. The above values, as well as many others, lie towards the center of histograms depicting arthropod densities in a variety of ecosystems (Petersen and Luxton 1982).

It is worth noting that within Mediterranean-type ecosystems, both numbers of individuals and species diversity generally decrease along the humid-xeric and natural-anthropogenic gradients (Bigot and Bodot 1973; Di Castri 1973; Karamaouna 1990). For instance, the characteristics of the composition of macroarthropod populations from a degraded evergreensclerophyllous system and a neighbouring phryganic one do not differ. Significantly lower numbers have been recorded in the latter and attributed to differences in primary production and the number of outcropping stones (Maggioris 1985). However, the species diversity of isopod communities in xeric habitats of Israel is higher than that estimated for more xeric Egyptian environments, but the opposite has also been cited. Warburg et al. (1978) recorded higher numbers of isopod species in rather xeric than in rather humid habitats of northern Israel.

In general, Acari – especially oribatid mites – dominate numerically among arthropods, while Collembola is the second most important group. Acari, followed by Collembola, are frequently also the most diverse groups (Postle et al. 1991). The ratios of oribatids to collembolan and of oribatids to prostigmatids have sometimes been used for rough characterisation of the physiognomy of the mesofauna in terrestrial ecosystems. The values of both ratios in Mediterranean ecosystems (Sgardelis et al. 1981) are lower than those reported for more xeric habitats (Wallwork 1972), while they are comparable with values reported for deciduous forests (e.g. Lebrun 1971).

Spiders and diplopods, followed by Coleoptera, are usually the most abundant surface and soil macroarthropods (Di Castri and Vitali-Di Castri 1981; Sgardelis and Margaris 1983, 1993; Karamaouna 1990). Moreover, there is evidence that within taxonomic groups only a few species dominate communities. For example, in a Greek insular ecosystem Trihas and Legakis (1991) reported that only two species, namely *Carterus calydonious* and *Acinopus subquadratus,* are by far the most dominant among Carabidae. Again, when studying isopods from mesic and xeric habitats in northern Israel, Warburg et al. (1984) found 15 species to be present, although the community was dominated by only two, namely *Metoponorthus pruinosus* and *Philoscia mus-*

corum. Analogous observations have been reported for the community of myriapods in sclerophyllous formations of Southern France (Bertrand et al. 1987) and for the diplopods of the evergreen-sclerophyllous formations of northern Greece (Iatrou and Stamou 1989a).

7.2
Seasonal Variations in Numbers

Seasonally varying numbers of Mediterranean arthropods have been documented in many cases (Paris 1963; Bigot and Bodot 1973; Maggioris 1985; Trihas and Legakis 1991, among others). However, seasonality in numbers is the outcome of a multifactor process, and direct relationships have scarcely been established between varying numbers and individual microclimatic or microhabitat components. For example, Warburg et al. (1984) were unable to establish definite relationships between isopod abundance and cover provided by rocks, stones or vegetation or between abundance and temperature or maximum, minimum or mean relative humidities. In general, such relationships are inferred indirectly. Again, Warburg et al. (1984) inferred a relationship between the pattern of precipitation and isopod phenology from the fact that, unlike the more xeric Egyptian habitats (Kheirallah 1980), numbers increased 1 month after onset of precipitation in xeric habitats of Israel.

Edmonds and Specht (1981) linked seasonality of numbers to seasonally varying resources. More specifically, in an Australian Mediterranean-type ecosystem two seasonal cycles were distinguished among plant consumers (Specht 1985). The first cycle, associated with a major response to temperature, follows variations in shoot growth and flowering of the dominant bushes, which occurs in late spring to summer. The second cycle, associated with a minor temperature response, is imposed by plant growth and flowering of the understory vegetation, which occurs in spring and is followed by summer inactivity. Furthermore, Specht (1985) suggested that a third annual cycle depending on temperature and moisture conditions in the litter and soil is imposed on decomposers.

The phenologies of soil arthropods from a Greek Mediterranean pine forest conform with the third type of cycle. These phenologies can be roughly divided into two groups. The first contains phenologies exhibited by more or less hygrophilous groups (holometabolan larvae, some pseudoscorpions, some diplopods, chilopods, and isopods) which are restricted during the rainy season, while the second is associated with less hygrophilous groups such as Araneae, Thysanoura, Dictyoptera, Hemiptera and Coleoptera (Karamaouna 1990). The temporal pattern of the latter is comparable to the patterns displayed by the macroarthropods of deciduous forests, while only the more hygrophilous groups exhibit seasonality.

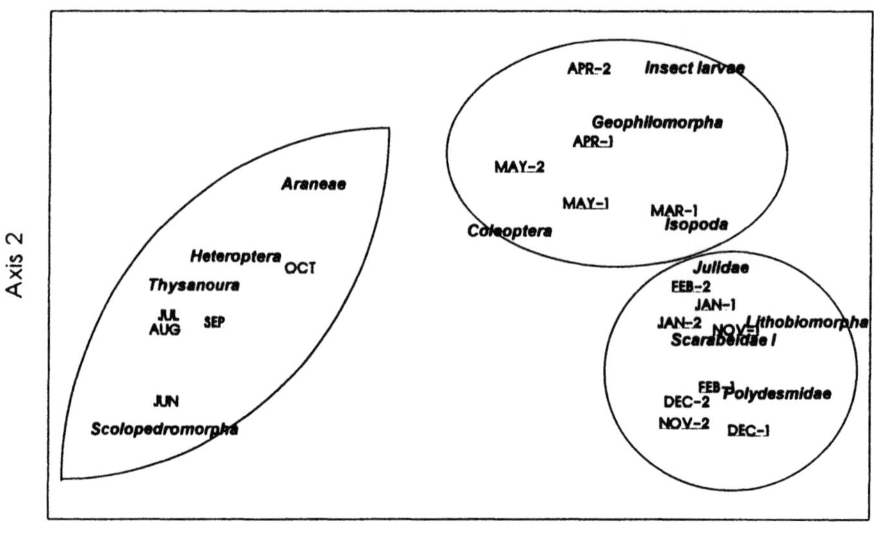

Fig. 7.1. Distribution of monthly samples of macroarthropod taxa on the plane of the first two axes of reciprocal averaging analysis (RA). (Sgardelis and Margaris 1983)

In more xeric habitats clear-cut seasonal patterns usually evolve. For example, three clearly distinguished periods of activity have been identified in soil macroarthropods from a phryganic ecosystem (Sgardelis and Margaris 1983). A wet period initiated immediately after the first autumnal rainfalls (November-February) is followed by a mild period (March-May), whereas the drought period lasts about 5 months (June-October; Fig. 7.1). Julidae are most abundant during the wet period, while Thysanura and Araneae are more numerous in summer. A sharp transition in the pattern occurs between spring and summer due to active avoidance of drought (Di Castri and Vitali Di Castri 1981). Indeed, by the end of spring, stadia that avoid drought are either replaced by drought-resistant ones or move into deeper layers. As a matter of fact, drought-susceptible insect larvae have for the most part reached adulthood by the end of spring, and adult stadia constitute the main bulk of drought-resistant arthropods. At the same time, drought-susceptible Julidae have already completed their migration to a depth of about 30 cm, where they aestivate in an inactive state (Fig. 7.2).

Analysis reveals further agreement among the activity patterns of macroarthropods and the factors driving litter decomposition (Sgardelis and Margaris 1983; Table 7.2). Likewise, the numbers of macroarthropods that decompose freshly fallen litter display a three-phase seasonal pattern. It is evident that autumnal rains trigger the onset of activity of both producers and macroarthropods. Considerable feeding activity results in the rapid for-

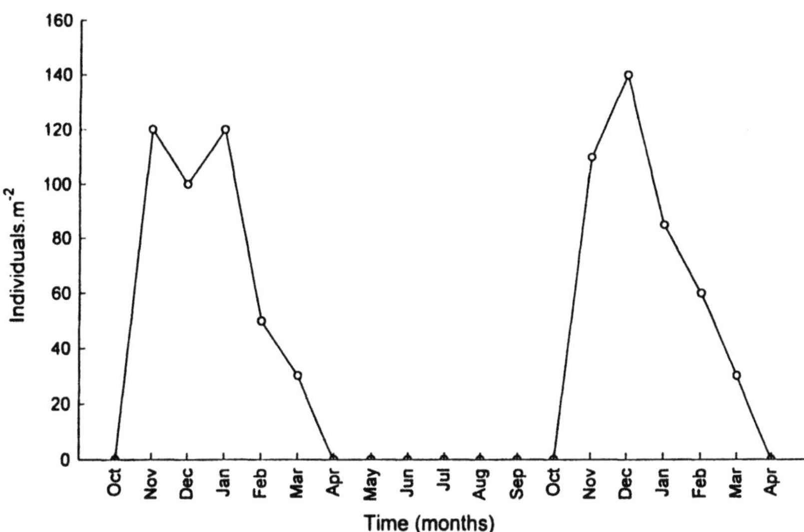

Fig. 7.2. Mean monthly numbers of Julidae in the litter samples of a phryganic ecosystem. (Data from Sgardelis and Margaris 1983)

Table 7.2. Climatic variables, decomposition correlates [a] and plant growth [b] in a phryganic ecosystem during the periods November-February (PI), March-May (PII) and June-October (PIII). (Sgardelis and Margaris 1983)

	PI	PII	PII
Average temperature ($^\circ$C)	11.67 ± 2.18	15.65 ± 2.68	24.92 ± 1.47
Minimum temperature ($^\circ$C)	8.42 ± 2.68	10.30 ± 3.52	20.39 ± 1.76
Maximum temperature ($^\circ$C)	15.24 ± 2.62	18.29 ± 4.15	29.02 ± 2.00
Mean rainfall (mm)	55.33 ± 16.14	32.28 ± 17.7	7.24 ± 5.66
Minimum rainfall (mm)	11.7 (Nov)	3.6 (May)	0 (all summer)
Maximum rainfall (mm)	91.2 (Jan)	54.4 (Mar)	23.2 (Oct)
Dehydrogenase activity (ng TPF g^{-1} soil)	0.237 ± 0.06	0.103 ± 0.06	0.095 ± 0.06
Nitrification potential (g NO^{-3} g^{-1} soil.day^{-1})	2.616 ± 0.22	1.95 ± 0.83	0.95 ± 0.45
NO^{-3} (g.g^{-1} soil)	25.3 ± 5.19	17.66 ± 1.99	22.46 ± 3.65
Plant relative growth rate (RGR)	3.62 ± 1.07	0.66 ± 1.94	-3.05 ± 4.4

[a] Data from Fouseki and Margaris (1981).
[b] Data from Margaris (1976).

mation of a superficial litter layer (3 cm in depth) consisting of plant detritus and animal excrement. As indicated by the presence of large amounts of dehydrogenase and considerable nitrification activity, this layer is rapidly colonised by microbes. Hence, autumnal activation of macrofauna induces the synchronous activation of the entire decomposition process, probably resulting in an increase in available nutrients during the growth period favourable to plants. The message from this example – as from many others – is that synchronised processes are the response of living organisms to seasonally varying Mediterranean environments.

7.3
Spatial Patterns

Due to small- and large-scale irregularities in the landscape, the bushy vegetation, the shallowness of the soil, and organic layers, as well as to anthropogenic impacts, the horizontal heterogeneity of Mediterranean habitats is higher than vertical heterogeneity. However, the vertical distribution of arthropods is crucial for their survival, especially during dry periods. Thus, the study of both the horizontal and vertical structures of soil communities is of primary importance for a thorough understanding of the dynamics of Mediterranean ecosystems.

7.3.1
Vertical Movement

Sgardelis and Margaris (1983) analysed data of mesofauna sampled in Mediterranean habitats using reciprocal averaging (RA). Although samples taken during the dry period were to some extent differentiated from wet period samples, the temporal pattern was rather complicated. Distortions were due to samples taken after an unusually heavy rainstorm in July and to artificially watered samples taken in September. These were incorrectly combined with samples taken in the wet period. To explain deviations from seasonality, the authors suggested that in response to summer drought, animals migrate to deeper soil layers where they either die after ovipositing or aestivate in an inactive stage. In all probability, increasing moisture stimulates egg hatching and/or reactivation of animals. Relatively high numbers can therefore be found in samples taken after heavy rain or artificial watering of the organic soil layers.

Due to low locomotory capacity, microarthropods rarely migrate (Metz 1971). By contrast, vertical movements have frequently been noted and are considered as ways to escape desiccation or high or low temperatures in the surface layers of the ground. Vertical movement to deeper soil is considered by Poinsot-Balaguer and Sadaka (1986) to be a mechanism allowing some Collembola to withstand summer drought. For example, during most of the year *Onychiurus zschokkei* is abundant in the intermediate litter layer of a Mediterranean *Quercus ilex* formation in southern France. However, at the beginning of the dry period some individuals move to the deeper layer to lay eggs. In the same vane, Ghabbour (1983) stated that most isopods in Egyptian habitats burrow to escape intensive heat and drought. He reported that individuals of *Porcellio olivieri* retreat to a depth of 10 to 25 cm. Paris offered an analogous explanation (1963) with respect to the spatial distribution of the isopod *Armadillidium vulgare*. In summer animals move downwards to soil as deep as 45 cm to avoid drought, while horizontal movement to moist microsites also offers refuge against drought. Animals emerge from their refuges at night to feed.

Vertical movements of epiphytic arthropods are related directly to climatic variables such as air temperature and relative humidity. For example, vertical movements of Araneae probably associated with seasonal changes in climatic variables were seen in bushes (Paraschi 1988). During summer in a xeric Greek ecosystem, most epiphytic Araneae were confined to a height below 70 cm, while during the wet period greater numbers were recorded within a range of 70-140 cm above the ground.

The above explanations link vertical migration directly to environmental conditions. Iatrou and Stamou (1991) correlated vertical distribution of the diplopod *Glomeris balcanica* with its foraging activity conditioned by the water content of the litter. In the wet Mediterranean formation on Hortiatis, *G. balcanica* is confined almost exclusively to the humus layer for most of the year, while during periods of highest feeding activity – from the middle of October to the end of November and again from the middle of March to the end of April – numbers are evenly distributed among the litter and humus layers.

7.3.2
Horizontal Distribution

The horizontal distribution pattern of most arthropod groups, such as pseudoscorpions, spiders and bristletails, varies seasonally (Magioris and Tsiourlis 1992). In general, small-scale horizontal patterns are associated either with intraspecifc relationships or with the distribution of sites for egg laying and sites of food resources, while large-scale patterns result from vectorial climatic, landscape and anthropogenic factors.

Vectorial factors condition habitat use by scorpions (Warburg et al. 1980). *Scorpio maurus fuscus* with its low potential for water regulation inhabits areas of high precipitation, dense vegetation and deep soil which provide suitable conditions for maintaining water and thermal balance in the deeper soil layers. By contrast, the efficient conserver of water, *Buthotus judaicus*, which is unable to dig borrows, is more abundant in rocky habitats which provide it with suitable microclimatic refuges.

Crawford et al. (1987) interpreted the prolonged dry-season foraging of *Archispirostreptus tumuliporus. judaicus* in the immediate vicinity of sheltered microsites as optimal exploitation of its microhabitat's potential for shelter. Living in environments characterised by high spatial heterogeneity, Mediterranean diplopods obviously share with desert species their high capacity to use microsites such as crevices, the nests of social insects, or soil burrows for shelter. Such behaviour involves horizontal aggregation, which is commonly observed in these animals (Crawford et al. 1987).

Joose (1971) related the capacity of Collembola to obtain shelter to the degree of moisture in the substrate. With increasing saturation deficit, locomotory activity increases. The more sensitive the species are, the more intense is their reaction to changes in saturation deficit. There is an inverse relationship between activity and the degree of aggregation. Species with high humidity requirements are forced to aggregate and therefore show less activity in the field.

Finally, it seems plausible that, in addition to abiotic factors, biotic factors such as circumscribed oviposition resulting in aggregations of eggs in favourable and protected microsites as well as food partitioning may also be important in determining the horizontal organisation of arthropod communities. In spatially uncertain Mediterranean habitats, phoresis, i.e. the attachment of a microarthropod to the outer surface of another arthropod for a limited time, might be an efficient mechanism for easy establishment in favourable microsites. Indeed, phoresis (which is of epigenetic origin) facilitates the dispersal of the individual or its progeny from unfavourable microsites (Athias-Binche 1995). However, phoresis in Mediterranean arthropods is seldom reported. Athias-Binche (1984) observed phoretic dispersal in mesostigmatids (Uropodinae). The phoretic is always deutonymph in that a specific fastening apparatus has been developed at the end of its body (Athias-Binche 1981, 1984). During travel the phoretic enters a latent life, ceasing to feed and develop. Activity is taken up again after detachment of the phoront and is induced by stimulus either from the carrier or the site of arrival. Mediterranean microarthropods exhibit rather facultative phoresis stimulated either by overcrowding or changing environmental conditions such as shortages of foods, air dryness etc. The carrier/phoretic relationship is infrequently species specific, and most phoronts can use a great variety of hosts (Athias-Binche 1995).

Finally, mass migration of macroarthropods among different habitat types is seldom observed (Warburg et al. 1984). Short displacements between adjacent microsites within single biotopes are also rarely reported (Trihas and Legakis 1991). Usually, no seasonal horizontal migrations are reported at all. For example, within a pine canopy, litter of *Pinus halepensis* and from under the understory shrubs of *Quercus coccifera*, *Phillyrea media* and *Arbutus unedo* was sampled for macroarthropods (Karamaouna 1990). No mass migration of animals from one litter type to another was detected.

However, in spatially heterogeneous habitats such as the Mediterranean one, populations usually split into interacting local populations with a finite lifetime, i.e. with a finite time of extinction (Hanski and Gilpin 1991). Accordingly, the maintenance of the genetic structure of such populations (favoured by dispersion of arthropods among patches) is of great importance for their persistence. Sheppard et al. (1992) studied the spatial distribution of oviposition in the sed predator *Larinus cynarae* (Curunlionidae). The arthropods mate almost exclusively on the capitula of the Mediterranean thistle *Onopordum illyricum illyricum*. In agreement with the general scheme described in previous paragraphs, the authors suggested that on a large spatial scale adults are able to distinguish patch quality for oviposition. Indeed, they established a relationship between capitula density (numbers of capitula/m²) in different patches and reproductive effort (numbers of eggs laid/capitula). However, within patches no correlation was found between microhabitat features such as plant size, local capitula density and oviposition level. Moreover, within patches reproductive effort is evenly distributed among different plants, irrespective of plant size or the availability of resources.

A mark-release-recapture experiment showed slight dispersion of arthropods among patches depending on patch size. However, even slight migration among patches (restricting risk) was effective in maintaining homogeneity in the genetic structure of the population. The overall conclusion is that *L. cynarae* living in a highly heterogeneous habitat maintains a rather homogeneous genetic structure and the population is not split into local populations. Slight dispersion either for mating or oviposition is of great ecological importance for the genetic stability of a population.

7.4
Community Structures Induced by Human Practices

Anthropogenic factors, namely fire and grazing, strongly affect the horizontal structure of Mediterranean ecosystems by increasing the heterogeneity of habitats. Depending on irregularities of the landscape, fire results in increased patchiness. In burned sites (local scale), both aboveground biomass and the litter beneath the canopies are completely destroyed, while intact vegetation patches restricted to narrow catenas remain. For a few months after fire, annuals, geophytes and resprouting shrubs cover the burned soil. They produce large quantities of litter, evenly distributed throughout the burned area. In contrast, grazing pressure results in the creation of paths and open sites between stands which are covered by grasses and soil lichens. Thus, grazing results in the creation of environmental gradients from the centers of shrubs towards exposed areas distant from them.

On a major scale, both fire and grazing result in increased horizontal heterogeneity of Mediterranean habitats. Resources are unevenly partitioned and the climatic conditions vary among different sites. Indeed, in exposed sites the available resources are limited and comparatively wider fluctuations occur in environmental variables. Consequently, the working hypothesis of relevant studies relates to the extent to which anthropogenic impacts lead to broader climatic and/or habitat selection on a major scale.

7.4.1
Fire-Induced Structures

Fire results in less heterogeneous and less detached habitats within burned sites. Consequently, reduced seasonal climatic variation as well as variation between microsites are expected. However, burned areas lose their capacity to buffer climatic fluctuations due to the loss of vegetation and litter which is expected to be stronger in burned sites (e.g. Lopes and da Gama 1994). Therefore, apart from the direct effect of fire, fire-induced changes in the availability of resources and in the structural and microclimatic correlates of habitats considerably affect both the numbers and composition of arthropod communities.

Referring to the direct effects of fire, Majer (1984), Saulnier and Athias-Binche (1986), as well as Lopes and da Gama (1994) reported a marked reduction in the numbers and biomass of epidaphic consumer and predator, while a minor reduction in the numbers of euedaphic ones was also recorded. The occurrence of fire during the active season has a considerable and immediate effect on fauna, whereas only a minor effect of fire is recorded at the end of the active period (Bornenisssza 1969 in Sgardelis and Margaris 1993; Majer 1984). Furthermore, the immediate effect of fire also depends on its intensity. Natural fires of high intensity have a destructive effect on numbers, whereas the effect of controlled burning is quite moderate. Finally, the effect of fire depends on the type of ecosystem. In general, fire is much less destructive in more xeric environments than in forests (De Izzara 1977). In the former, increased populations have even been recorded shortly after fires (Pomeroy and Rwakaikara 1975). According to Majer (1984), soil and litter invertebrates exhibit a broad array of responses to fire, including immediate reduction in numbers, delayed reduction in population densities, and temporary absence of some taxa while others remain unaffected by fire or may increase their numbers.

7.4.1.1
The Effect of Fire on Microarthropods

Short- and long-term effects of fire on Mediterranean micro- and macro-arthropods have been studied by Sgardelis (1988), Sgardelis and Margaris (1993), and Sgardelis et al. (1995). Results concerning the numerical response of some soil microarthropods are summarised in Table 7.3. The first finding is that the immediate effect of fire is most pronounced on litter dwellers, while animals that live under stones remain almost unaffected (Figs. 7.3, 7.4).

Table 7.3. Numerical responses to fire of some soil dwelling microarthropods of a phryganic ecosystem. (Sgardelis 1988)

Taxonomic group	1st year	2nd year
Litter		
Mature and immature oribatids: *Scheloriates latipes* and *Chamobates* sp, Bdelidae, mesostigmatids (except Rhodacarida)	↘	→
Oribatids (*Hypochthonius* sp. and Oppiidae), Rhodacaridae	→	↘
Collembolans	↘	↗
Scutacaridae	⇔	⇔
Stones		
Most taxa	⇔	⇔
Scheloribates latipes	↗	→
Collembola: arthropleones	⇔	↗

↘: decreasing numbers
↗: increasing numbers
→: constant numbers
⇔: unaffected

Samples from microsite litters display the greatest variability and are positioned towards the endpoints of the first axis of the canonical variation analysis (CVA) graph. Moreover, the preponderance of seasonal variability is obvious, and samples from the wet period are grouped towards the right endpoint of the first axis, while those from the dry period are grouped near to the left end. The variability detected between unburned and burned litter, although lower than seasonal variability, is nevertheless remarkable.

On the second axis of the CVA graph, less pronounced differences in numbers sampled under stones in unburned and burned sites are depicted. This is due to the positive effect of fire on certain groups such as some collembolans and the oribatid *Scheloribates latipes*. Normally inhabiting litter, these animals display broad habitat tolerance and may also exploit resources under stones on the ground surface during the first year after fire.

Finally, most Acari and collembolans are more abundant in unburned litters during the second year due to more favourable moisture conditions there. However, contrary to other microarthropods, this trend (indicating recovery capability) was not recorded among Acari and Collembola in burned litter. An interannual trend was also negligible in both burned and unburned sites among animals that shelter under stones.

Although seasonal and interannual effects predominate over those induced by fire, natural and deliberate fires occurring every 3-5 years have an overriding impact on the structure and dynamics of Mediterranean ecosystems and are a very important selective factor for microarthropods. Mechanisms that allow for the survival of an adequate number of animals, coupled

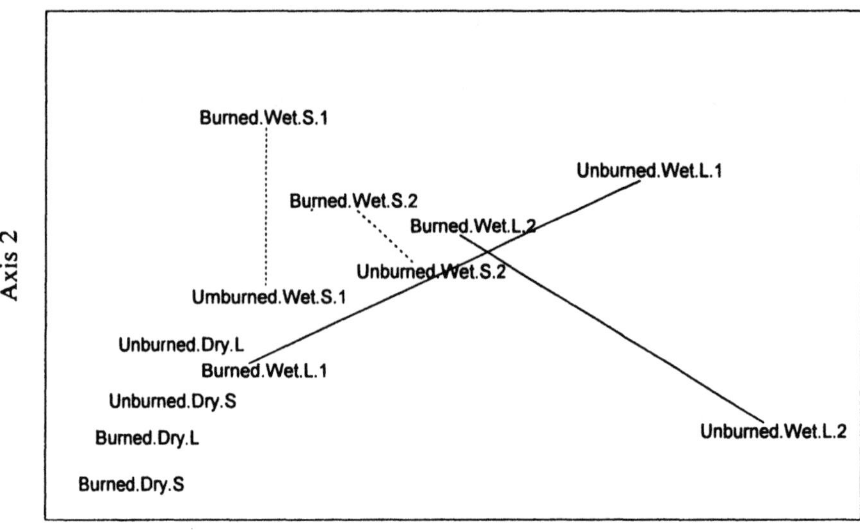

Fig. 7.3. Canonical variate analysis CVA graph for the discrimination of litter and soil samples from a phryganic ecosystem grouped by *site (Unburned, Burned)*, microsite (*L* litter;, *S* stones), season (*Wet, Dry*) and year (*1* first postfire; *2* second postfire). *Lines* indicate the effect of the fire (litter is represented by a *solid line* and stones by a *dashed line*). The analysis was performed on the abundance of 16 taxa of Acari (Sgardelis and Margaris 1993)

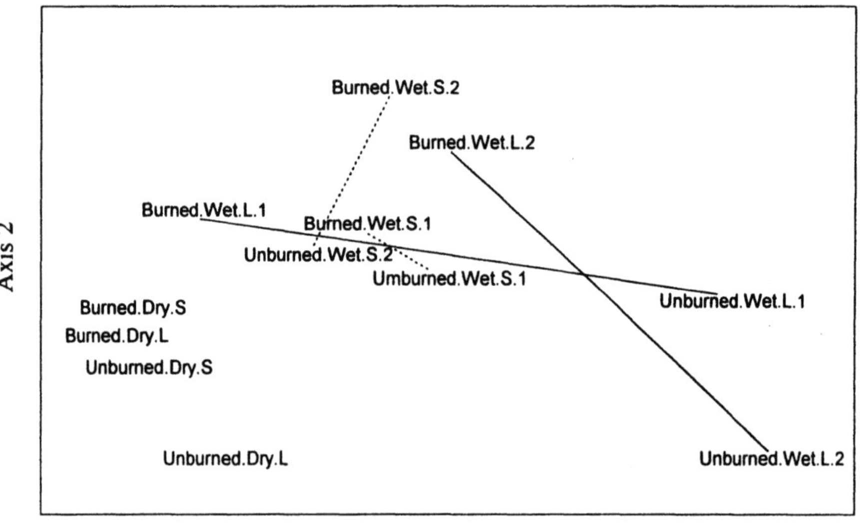

Fig. 7.4. Canonical variate analysis CVA graph for the discrimination of litter and soil samples from a phryganic ecosystem grouped by *site (Unburned, Burned)*, microsite (*L* litter; *S* stones), season (*Wet, Dry*) and year (*1* first postfire, *2* second postfire). *Lines* indicate the effect of the fire (litter is represented by a *solid line* and stones by a *dashed line*). The analysis was performed on the abundance of 13 taxa of microarthropods (Sgardelis and Margaris 1993)

with capabilities for rapid recolonisation of destroyed areas are essential components of the adaptive strategies of microarthropods. The adaptations of litter-dwelling animals can be illustrated by collembolans and oribatids. Collembolans are strongly reduced in numbers after fire. They can, however, recover rapidly thanks to species able to colonise a wide range of habitats that survive in sheltered sites under stones. Similarly, most oribatids resist the immediate effect of fire, while species occupying a wide range of habitats such as *S. latipes*, can exploit intact resources under stones lying on the surface of the soil. Fire apparently engenders the selection of broad habitats, increased reproductive effort, and rapid development. In general, burned areas are dominated by species such as the uropodid mite *Olodiscus minimus* (Athias-Binche 1987) that select a wide range of habitats and climatic conditions.

From supplementary data collected annually, Sgardelis and Margaris (1993) estimated that the recovery of microarthropods in the Greek phryganic ecosystem lasts roughly 3-4 years. Analogous durations were also reported Majer (1984) for a Western Australian Mediterranean-type forest, while Lopes and da Gama (1994) recorded even higher densities of Collembola accompanied by a slight decrease in the number of species. Apparently, the recovery of microarthropods follows the rapid recovery of Mediterranean vegetation. According to Athias-Binche (1987), recolonisation is due either to the migration of animals that take refuge in the deeper soil layers or to facultative phoretics such as *O. minimus* which are the most resistant and dynamic colonisers.

7.4.1.2
The Effect of Fire on Macroarthropods

7.4.1.2.1
Numerical Response

The intensity of either natural or deliberate fires is usually moderate in Mediterranean ecosystems. Data provided by Abbott (1984), Majer (1984), and Saulnier and Athias-Binche (1986) show that, due to the decline in food resources, saprophagous litter species are most susceptible to this kind of fire. Thus, the main constraint on the recovery of long-lived arthropods such as julids is the shortage of resources.

The effects of a fire that took place in August 1980 in a Greek phryganic ecosystem were studied by Sgardelis et al. (1995). The fire was of moderate intensity and its immediate effect on macroarthropods was negligible, since most of them were absent from the topmost organic layers at the time of fire. Thus, unlike the case for the majority of microarthropods, the long-term effect of fire on macroarthropods is more salient than the short-term one, as might be expected.

Table 7.4. Numerical responses to fire of some soil dwelling macroarthropods of a phryganic ecosystem. (Sgardelis 1988)

Taxonomic group	1st year	2nd year
Julidae	↘	→
Insect larvae	↘	↗
Coleoptera	⇔	↘
Isopoda, Geophilomorpha, Aranae	⇔	⇔

↘: decreasing numbers
↗: increasing numbers
→: constant numbers
⇔: unaffected

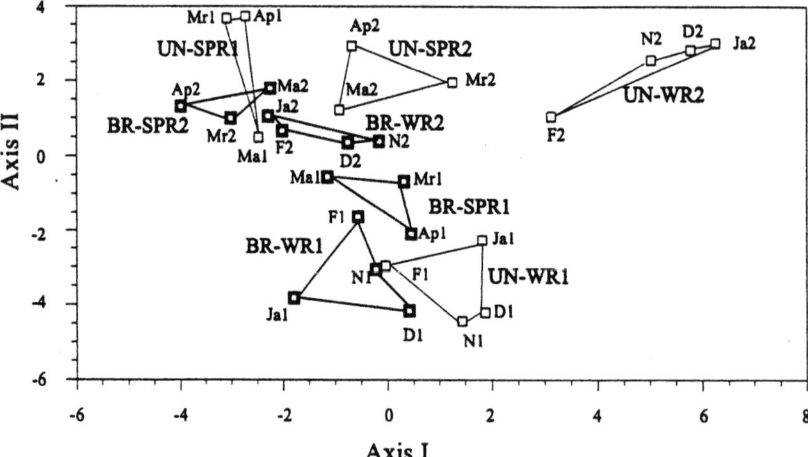

Fig 7.5. CVA two axes plane for the discrimination of monthly sample groups prefixed by site (*UN* unburned; *BR* burned), season (*WR* winter; *SPR* spring), year (*1* first postfire; *2* second postfire) and months (*Ja-D* January-December). Monthly samples of each group are connected by *line segments* (Sgardelis et al. 1995)

The effect of fire on macroarthropod groups is shown in Table 7.4, while Fig. 7.5 shows the results of a CVA analysis on the plane of the first two axes (summer samples were excluded due to the absence of arthropods). As with microarthropods, seasonal variability in the unburned site was greater than interannual variation or that induced by fire. Interannual variability was greater in the burned site, while during in the first postfire year the effect of fire was most obvious in spring.

Based on the above results, Sgardelis et al. (1995) made the following suggestion: although obvious – due to depressed numbers of julids – the effect of fire is not pronounced 2.5 to 5.5 months after burning (November-February). Numbers of julids continue to decline from winter to spring, while insect

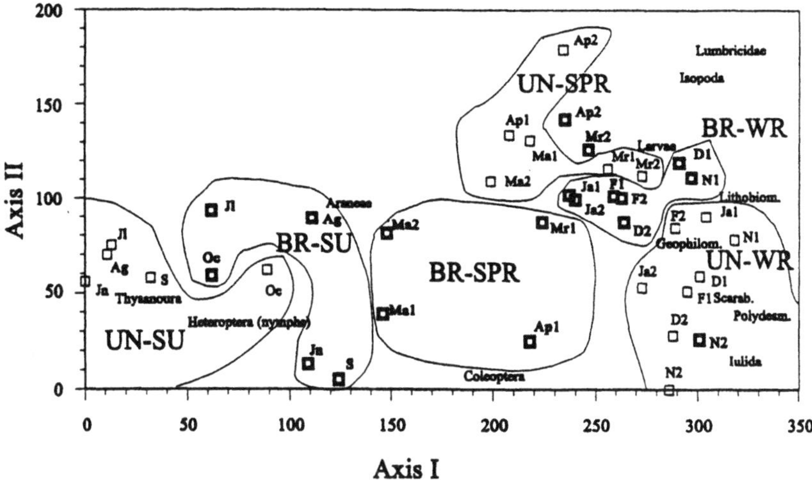

Fig. 7.6. Distribution of monthly samples and macroinvertebrate taxa. Groups of samples from the same site (*UN* unburned; *BR* burned), season (*WR* winter; *SPR* spring; *SU* summer) and months (*Ja-D* January-December) are enclosed. *Numbers 1,2* denote the postfire year of sampling. (Sgardelis et al. 1995)

larvae and Araneae increase considerably in numbers. Dissimilarity between sites declined 19 months after fire due to the recovery of insect larvae, while julids were sampled again and found to be sparse in number in the burned site. Finally, the delayed decline in numbers of Coleoptera was attributed to climatic constraints rather than to the effect of fire.

7.4.1.2.2
Community Composition

To examine fire-induced changes in community composition, Sgardelis et al. (1995) exploited data using multivariate techniques and concluded the following: in spite of fire-induced decline in the proportion of Julida and Coleoptera, the ranking of these groups in spring did not change significantly. In the dry period, however, the effect of fire on group ranking is considerable due to dominance of coleopterans over Thysanoura.

In Fig. 7.6 the results of detrended correspondence analysis (DCA) are graphed on the plane of the first axis. The seasonality of the monthly samples is depicted on the first axis of the DCA graph. The winter samples (including diplopods, chilopods and scarabeiform larvae) are grouped closer to the right endpoint of the axis, spring samples occupy the middle of the axis, while summer samples including Thysanoura and Heteroptera occupy its left end. Fire diminishes the effects of seasonality by shifting the records of both summer and winter months towards the middle of the first axis. It is evident that changes in community structure are chiefly driven by seasonally varying climatic variables both in the unburned and the burned site. However, due to the effect of fire on the structural properties of the habitats and the accompanying changes in climatic variables, the taxa in the burned site are more evenly distributed throughout the year.

As has been shown, the long-term effect of fire on both numbers and community structure within regular seasonal changes is due to a reduction in aboveground vegetation cover and litter, which in turn results in changes in microclimatic conditions and food resources. Thus, no specific adaptations of habitat selection are anticipated since the numerical recovery of fauna as well as the restoration of the various communities depend on the recovery of vegetation. In contrast, fire leads to broad climatic selection by inducing stronger fluctuations in environmental variables. Indeed, species resistant to sudden changes in their environment, such as diplopods, are equally good colonisers of destroyed areas. In fact, diplopods (e.g. some Glomeridae and Julidae) tolerate temperature changes in a range from -3 to about 50 °C, whilst, they can resist xeric conditions by conglobating. Although the overall reproductive potential of diplopods is low, under favourable climatic conditions major recruitment to the population occurs every 3-4 years, corresponding with interannual climatic cycles. Long-lived diplopods evidently confront periods of limited resources after fire with their low numbers, whereas population recovery follows restoration of the overstory vegetation and accompanying improvements in microclimatic and food conditions after 3-4 years.

Reconstruction of aerial fauna takes place faster. Giliomee (1989, 1992) recorded almost total destruction of aerial fauna in a fynbos after fire. Nevertheless, 7 months later 11 out of 12 species of grasshoppers had returned, while five supplementary species were also recorded.

7.4.2
Grazing-Induced Structures

The organisation of oribatid and collembolan communities in a maquis formation on Mt Hortiatis was investigated by Asikidis and Stamou (1991) and by Argyropoulou et al. (1994) respectively. Figure 7.7 shows the arrangement of monthly samples of collembolan species on the plane of the first two axes of a DCA biplot. Species composition changes drastically before and after the dry period, and community structure in the dry period is entirely different from that in the wet period. The seasonal effect on numbers of collembolan species is so pronounced as to lead to bipolarity in community composition rather than to a gradient in time. It is again evident that changes in numbers as well as horizontal organisation of the communities due to overgrazing lie within the seasonal boundaries.

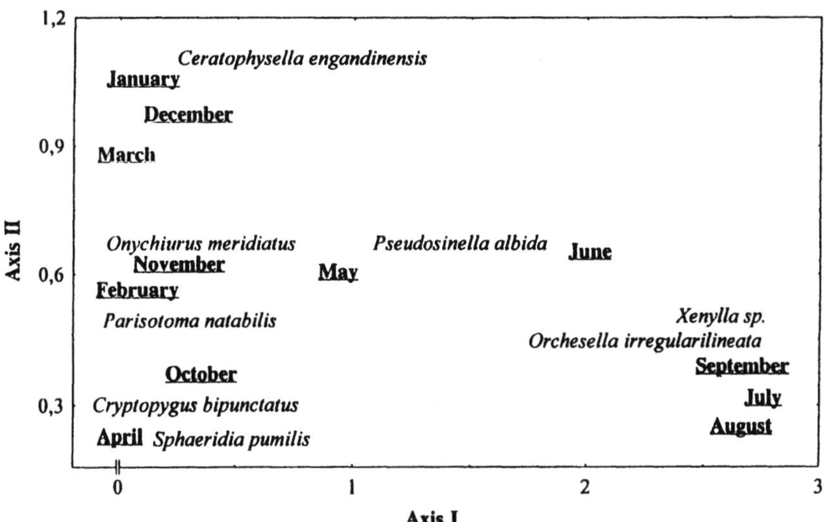

Fig. 7.7. Distribution of monthly samples of collembolan species on the plane of the first two axes of DCA. (Argyropoulou et al. 1994)

Overgrazing significantly affects numbers. Indeed, mean monthly density recorded in exposed or partly sheltered microsites are comparable to those recorded in arid ecosystems, whereas higher average densities comparable with those reported in deciduous forests were recorded in sheltered microsites. The community structure is shown in Fig. 7.8, while in Fig. 7.9 the distribution of microsites is indicated along with that of the main collembolan species with respect to their preference for varied microsites. The epiedaphic species changes successionally from *Isotomurus palustris* to *Pseudosinella albida*, and an analogous successional change is detected for the euedaphic species from *Onychiurus meridiatus* to *Xenylla* sp. The only species exhibiting discontinuous spatial distribution are *Orchesella irregularilineata* and *Ceratophysella engadinensis*. Data analysis shows that, like fire, grazing, although of less importance than seasonality, is again an important selective force resulting in broad habitat selection.

To explore the horizontal structure of a collembolan community with respect to habitat selection, Argyropoulou et al. (1994) tested the core-satellite hypothesis on animals sampled on Hortiatis (Hanski 1982, 1991; Fig. 7.10). The results of the analysis show that in addition to the succession of species along the exposed-sheltered microsite gradient, changes also occur at the community level. In the open sites species fall either into the "core" category occurring in large assemblages or into the "satellite" category forming small assemblages. Apparently, the spatial organisation of collembolan communi-

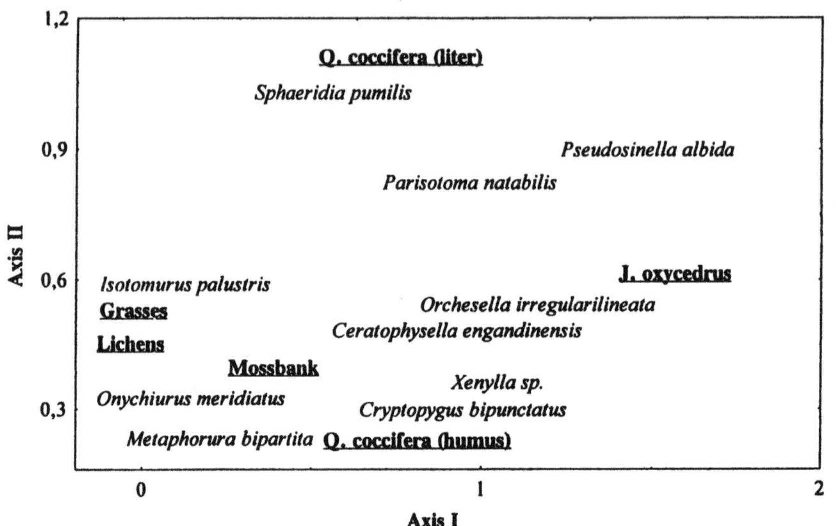

Fig. 7.8. Distribution of microsites and collembolan species on the plane of the first two axes of DCA. (Argyropoulou et al. 1994)

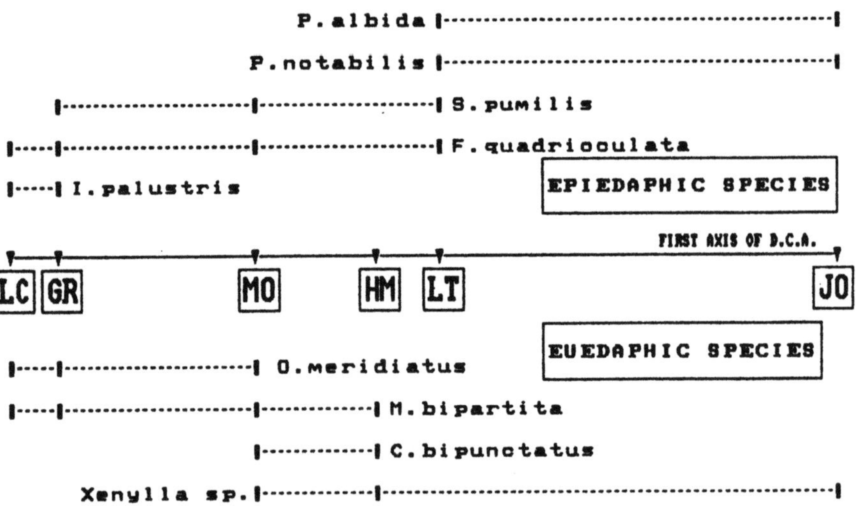

Fig. 7.9. Pattern of succession of epiedaphic and euedaphic collembolan species along the environmental gradient depicted on the first axis of DCA. *JO* Litter and humic horizons of *Juniperus oxycedrus*; *HM* humic horizon *of Quercus coccifera*; *LT* litter of *Q. coccifera*; *MO* moss banks; *GR* patches of Graminae species; *LC* patches of soil lichens. (Argyropoulou et al. 1994)

ties at these sites is related to vectorial abiotic factors rather than to resource partitioning or predation. In exposed sites annual fluctuations in climatic variables are large, consequently core species such as the broad temperature-

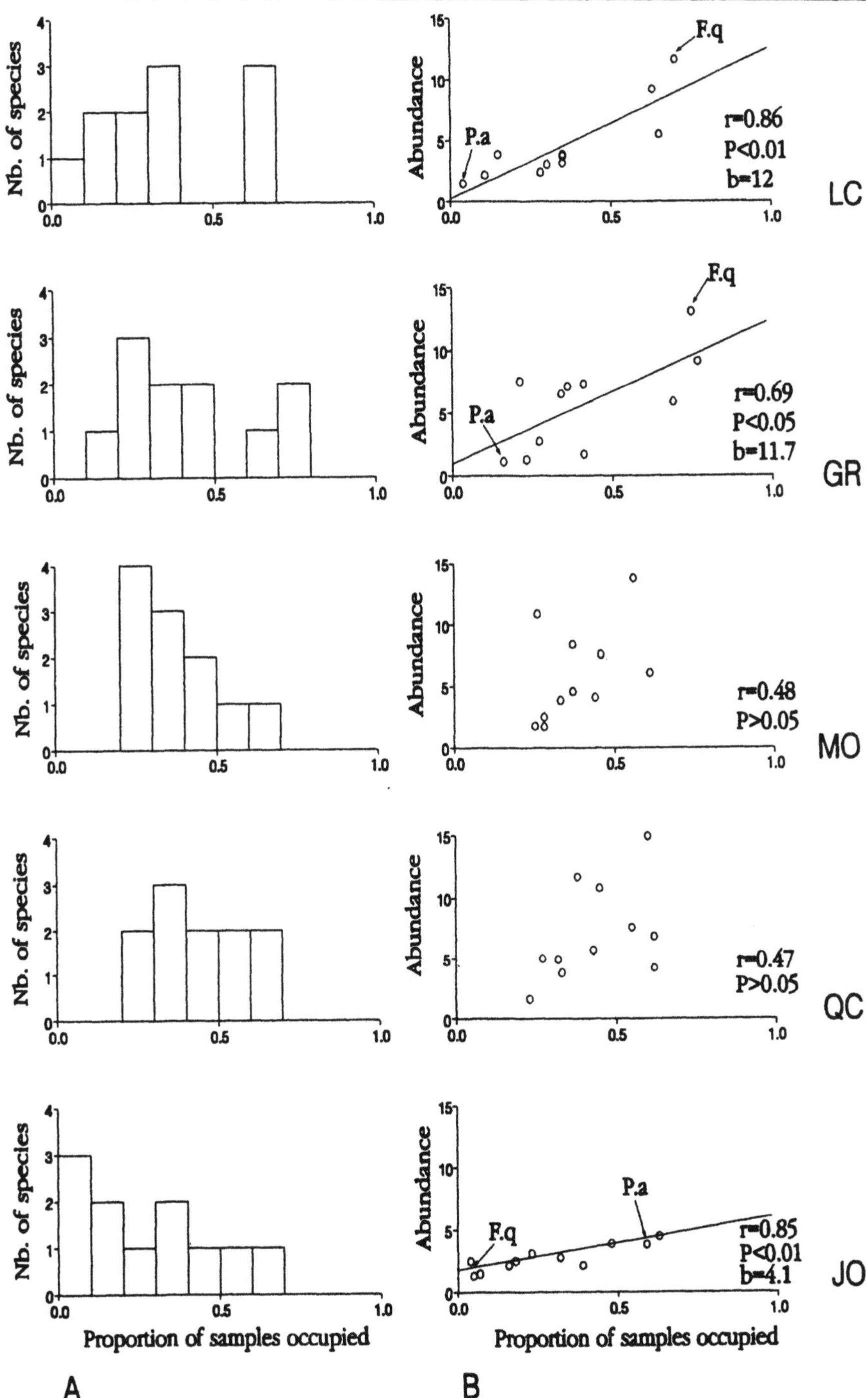

Fig. 7.10. A Distribution of 11 collembolan species among proportions of occupied samples within different microsites (abbreviations of microsites as in Fig. 7.9). B Relationship between distribution, i.e proportion of samples occupied by the species, and abundance, i.e. mean numbers in the occupied samples, for 11 collembolan species. *Arrows* indicate examples of core-satellite switching; *P.a Pseudocinella albida; F.q Folsomia quadriculata.* (Argyropoulou et al. 1994)

selected *O. meridiatus* (Chap. 3) can dominate in the satellite category. It seems plausible that physiological constraints may underlie habitat selection in exposed sites. In contrast, the community of sheltered sites appears homogeneous and numbers are evenly distributed among species in sheltered sites. In these communities, where moderate climatic fluctuations occur, the effect of the biotic components of the environment is, as anticipated, higher. It can be observed that core species of open sites such as *Folsomia quadrioculata* and *O. meridiatus* switch to satellite status in sheltered sites. Finally, in the outer site no distinction between core and satellite species has been made, while only the relationship between abundance and distribution is linear. In accordance with Maurer (1990), it can be concluded that habitat in this site is relatively simple with few resources. Argyropoulou et al (1994) suggested that the partitioning of limited food resources probably caused spatial separation of niches in simple habitats.

7.5
The Biotic Correlates of Habitat Selection

Although no detailed data are available on the feeding habits of arthropods, most Mediterranean species are considered generalists. For example, Crawford et al. (1987) reported that no food specialisation was observed among diplopods. Paris (1963) reported that, although the isopod *A. vulgare* shows a net preference for the decaying leaves of *Picris echioides*, population density and age structure in the field is independent of the distribution of plant debris. He suggested that this species, which is broad food-selected, can easily spread and establish itself in new sites. In contrast, habitat selection among oribatids within single shrubs is linked to their feeding habits (Sgardelis et al. 1981; Fig. 7.11). Unspecialised feeders and panphytophages according to classifications by Berthet (1964) and Luxton (1972) are confined to the center of the shrubs, while microphytophages and specialised feeders prefer the outer borders.

Food correlates of habitat selection have been further explored in laboratory cultures of the oribatid *Scheloribates* cf. *latipes*. In Table 7.5 the effect of diet on demographic parameters is shown. Feeding on lichens as well as on decomposing litter enables the life cycle development of *C.* cf. *latipes* to take place. Higher fecundity, lower mortality and lower CV% values were recorded in cultures containing the lichen *Xanthoria parietina*, decomposing litter and mosses. It must be noted that the latter are not palatable but constitute sites for egg laying and preecdysial immobilisation of juveniles. Analogous nutritional preferences were exhibited by all other oribatid species tested (Asikidis, unpublished) and by the majority of collembolan species (Argyropoulou, unpubl.). It seems that epiphytic and soil lichens, whose abundance is related to the intensity of grazing pressure, are essential for the life cycle develop-

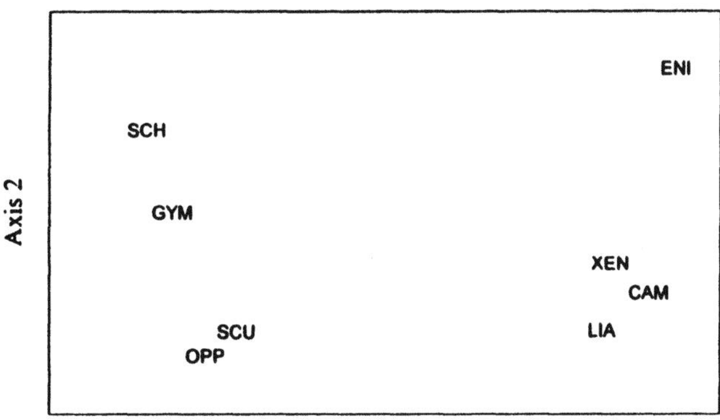

Fig. 7.11. Distribution of the oribatid mites sampled within the canopies of *Thymus capitatus* on the plane of the first two axes of principal components analysis (PCA). *ENI Eniochthonius minutissimus; XEN Xenillus tegeocranus; CAM Camisia* sp.; *LIA Liacarus coracinus; SCH Scheloribates latipes; GYM Gymnodamaeus bicostatus; SCU Scurtovertex* cf. *bulgaricus; OPP* Oppiidae. (Modified from Sgardelis et al. 1981)

Table 7.5. Effect of diet on the fecundity (f_x) and mortality (m_x) of *Scheloribates* cf *latipes* (Oribatidae) with CV% the coefficient of variation. (Stamou and Asikidis 1992)

Diet	f_x	CV%	m_x	CV%
X. parietina	0.056 ± 0,010	38	0.74±0,10	25
P. adscedens	0.057 ± 0.020	66	0.66±0.30	66
C. foliacea	0.003 ± 0.001	22	1.58±0.10	19
Moss	-	-	3.30	-
Fresh litter	-	-	3.10±0.20	10
Dried litter	0.004 ± 0.004	73	1.68±0.06	7
Decomposing litter	0.044 ± 0.020	78	1.28±0.20	31
G. balcanica faeces	-	-	2.90±0.30	13
X. parietina+decomposing litter+Moss	0.088 ± 0.010	13	0.40±0.04	16

ment of microarthropods on Hortiatis. Furthermore, taking the subtle nutritional preferences of the dominant macroarthropod *G. balcanica* into account (Iatrou 1989), it can be inferred that high food specialisation characterises the arthropods that inhabit the degraded Mediterranean ecosystem on Hortiatis. In addition, the contamination of food and substrate by deposited excrement and developing hyphae results in a pronounced decline of the demographic parameters of arthropods. Survivorship and fecundity of the oribatid *S.* cf. *latipes* were drastically depressed in cultures in which food was replaced and excrement removed every 15 days compared with cultures in which food was changed and excrement removed every 3 days (Fig. 7.12). In general, food quality also directly affects the duration of development of

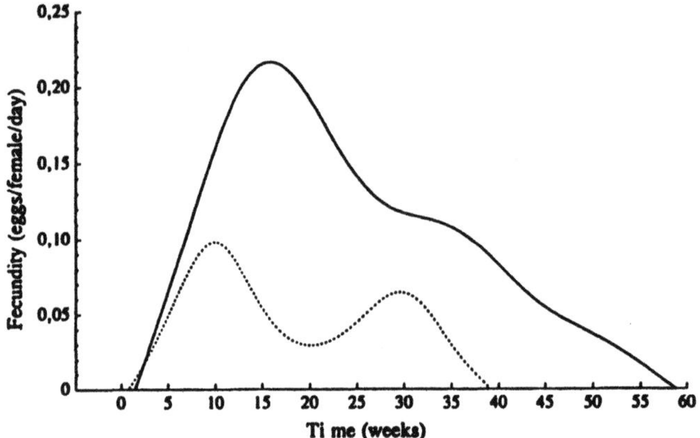

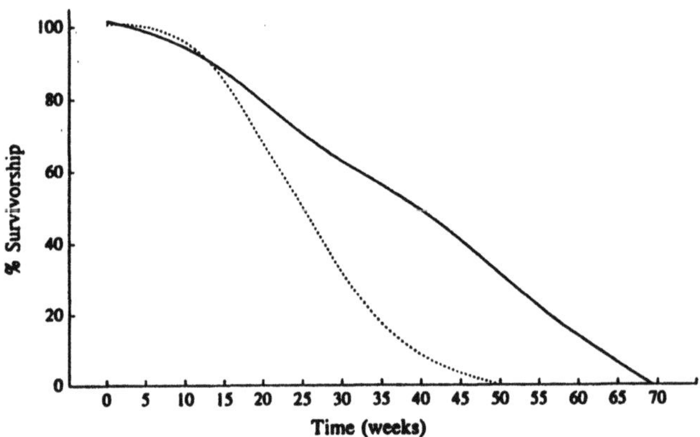

Fig. 7.12a,b. Effects of the quality of food and the culture substrate on fecundity and survivorship of the oribatid *S.* cf. *latipes*. *Solid lines* represent food renewed and faeces removed every 15 days, *dashed lines* food renewed and faeces removed every 3 days. Data were fitted by least squares. (Data from Asikidis 1989)

arthropods (Stefaniac and Scheniczak 1981; Norton 1994). Consequently, if most Mediterranean arthropods are food specialists unable to shift in their food resources, then food quality might also be a selective force.

Baker (1979b) reported changes with age in habitat selection of the millipede *Australiosoma castaneum*. They were attributable partly to feeding preferences and to predation. Predation was also studied in the oribatid *S.* cf. *latipes*, and data show that juveniles are subject to intensive predation by mites of the family Bdelidae. The predation rate increases with prey density

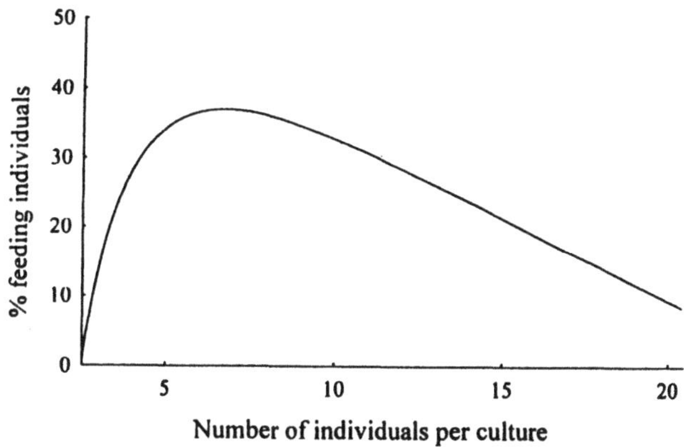

Fig. 7.13. "Allee-type" density dependence of the aggregation size of specimens of S. cf. *latipes* on a food item. (Data from Asikidis 1989)

and predator response falls in the "Type-A" or "Type-B" categories of Holling (1959). Subsequent instars appear more vulnerable to predation, and searching efficiency of the predators increases when the subsequent instars of the oribatids constitute the prey. Presumably, earlier instars of S. cf *latipes* escape predators by taking refuge in microscopic spaces of the culture substrate. No attacks on immature stadia in preecdysial immobilisation or on adults were recorded. Consequently, only about two thirds of immature mites and no adults were vulnerable to predation. In general, most arthropods are considered to be vulnerable to a wide array of predaceous species, and features such as the sclerotisation of adults or long adult survival can be viewed as adaptations to predator defense. Unlike predation, disease, parasitism and cannibalism are negligible causes of death (e.g. Paris 1963; Baker 1979b, 1985).

The biotic correlates of habitat selection in arthropods from other Mediterranean regions have not yet been studied. However, from the information gathered on Hortiatis it can be concluded that food composition and quality, the quality of the substrate, and infrastructure of the microhabitat (the presence of fresh moss leaves, microscopic holes etc.) coupled with heavy predation are factors that restrict habitat use and strongly affect the demography of the fauna. Mechanisms allowing animals to withstand the severity of the biotic environment are explored in some Acari and collembolan species (Stamou and Asikidis 1992; Argyropoulou, unpubl.). The rate of displacement and the ability of animals for aggregative food exploitation are density dependent. Moreover, the positive effect of density is of greater importance at lower densities (Fig. 7.13). Laboratory data show that the aggregation of animals on their food is not due to active attraction. Food searching activity is random and density dependent. Only a few animals can aggregate and exploit individual food items. It seems that contact between specimens stimulates

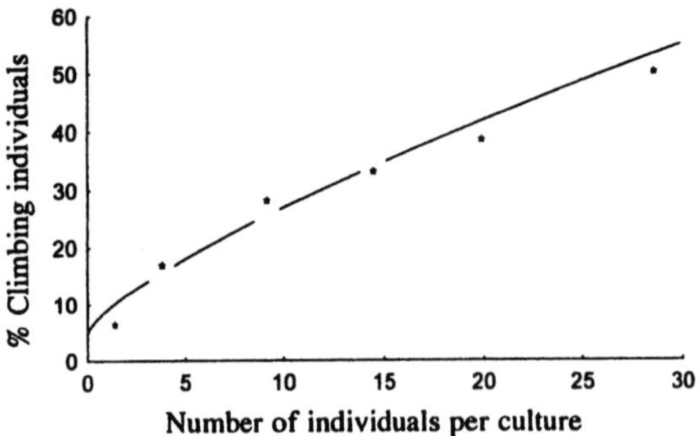

Fig. 7.14. Relationship of specimens of *S.* cf. *latipes* forced to abandon the culture substrate on account of population density. (Data from Asikidis 1989)

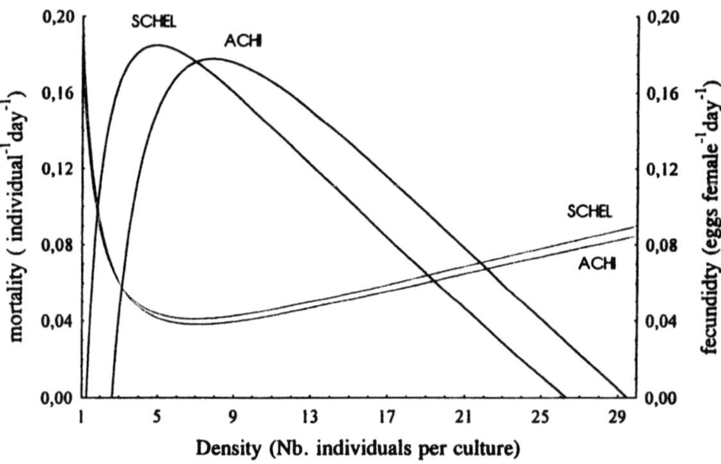

Fig 7.15. "Allee-type" density dependence of fecundity and survivorship in the oribatids *S.* cf. *latipes* (*dashed lines*) and *A. oudemansi* (*solid lines*). (Data from Asikidis 1989)

them and increases their mobility rate. Consequently, in crowded cultures most of the animals abandon the substrate and climb the walls of the vessel containing it, presumably to avoid contact with other individuals (Fig. 7.14). Moreover, density-dependent fecundity and survivorship fall into the "Allee type" category of Fujita (1954). In Fig. 7.15 the density-dependent demography of some selected species has been graphed. Up to an optimal threshold, increasing density has a positive effect upon demographic parameters such as fecundity and survivorship, whereas beyond this threshold the effect of increasing density is negative.

Due to trade-offs between attraction and repugnance, soil arthropods can apparently develop spatially distributed density-dependent mechanisms. These are probably also related to weather conditions (Paris 1963) and regulate population sizes within microsites. The formation of small aggregations in favourable microsites results from attraction, while a density-dependent rate of dispersion increases food searching efficiency and protects resources from overexploitation and contamination. It also diminishes the searching efficiency of predators. Specialisation allows animals to partition food resources in more or less homogeneous sheltered sites, while density-dependent behaviour favours the exploitation of food. The overall conclusion is that arthropods with highly specialised diets may achieve optimal habitat selection at a local population level due to their capability for aggregating on food items coupled with "Allee-type" density-dependent demography.

CHAPTER 8

Synthesis

The environment of the Mediterranean-type ecosystems appears spatially fragmentous and highly heterogeneous in time. Temporal heterogeneity resides upon strong interannual, seasonal and diurnal oscillations of climatic variables inducing analogous fluctuations in food and water resources. The spatial heterogeneity of the Mediterranean habitats is to a lesser extent originated from the heterogeneity of the landscapes but for the most part from management practices involving overgrazing of the lands coupled with frequent fires and wood removal. It is evident, however, that seasonal effects are most important and overwhelm effects based either on interannual fluctuations or on ordinary human impacts.

The combined effect of climatic and human impacts on Mediterranean ecosystems results in the continuous degradation of the Mediterranean lands. Indeed, climatic variation, intensive grazing, fire cycles and wood removal have degraded or drastically changed the natural Mediterranean vegetation, resulting in fine-grained mosaics which represent degradation stages and induce unevenly distributed resources. Mediterranean regions are characterized by the co-existence of diversified formations from the sub-humid to the xeric, and the severity of their environment is based upon both the strong temporal oscillations of the climatic variables and the fragmentary structure of habitats. However, the effect of both climatic and human impacts appears to a great extent predictable, although unpredictable variation can also be recorded.

As has been shown (Stamou 1995; Stamou and Pantis 1995), plant and animal species have adopted different kinds of solutions to overcome predictable and unpredictable unfavourable periods in their environment. Notwithstanding severe environments, plant and animal adaptations result in high system resilience which ensures the persistence of the Mediterranean-type formations for millennia. Thus, within a context involving counteracting forces such as strong climatic and anthropogenic impacts leading to degradation on the one hand and resisting organisms on the other, the goal of this book was to describe strategies shown by arthropods in response to the severity of the Mediterranean habitats.

Initiated by Cole (1954), strategic thinking in ecology resides upon the concepts of allocation and strategy (Korfiatis and Stamou 1994). The corner hypothesis underlying the development of schemata describing life history strategies is rather simple: a limited amount of time and energy is available throughout organisms' life. Hence, limited quantities should be invested in various life cycle attributes in such a way as to achieve optimal adaptation of organisms to environmental constraints. Obviously, the timing of life cycles and the patterns of energy allocation drive the development of adaptive strategies, i.e. the development of those life history traits which maximise the fitness of organisms. Thus, adaptive strategies are viewed as responses of organisms to genetic and environmental constraints resulting in trade-offs among life history characteristics.

The life history model most familiar to ecologists is that of r-K selection. First introduced by MacArthur and Wilson in 1967, this model captured the concept of life history strategies and overwhelmed strategic thinking until the late 1970s. Pianka (1970) introduced the schema of the r-K line continuum on which species can be classified according to the number of produced eggs and the survival of juveniles. According to model predictions, near the r extreme of the line continuum species can be classified which evolved under density-independent conditions (r-strategists). r-strategists are short-lived and display high reproductive potential. By contrast, K-strategists (species classified closer to the K endpoint of the r-K gradient) evolved under density-dependent conditions, are long-lived and display low reproductive potential.

Strongly criticising the r-K selection model for its simplicity, Greenslade (1972a,b), Southwood (1977, 1988) and Grime (1977, 1979) introduced habitat-templet models. More specifically, Grime and Greenslade added to the original r-K line continuum a supplementary selection gradient termed "stress" or "adversity selection", while Southwood stressed the significance of habitat structure as a templet for ecological strategies. Southwood et al. (1974) based upon cost-benefit ideas originated the idea that the structural characteristics of habitat make up a templet against which evolutionary pressures fashion the ecological strategy of a species. This means that the selection of life history strategies can be achieved only in terms of environmental variables (Korfiatis and Stamou, submitted).

The novelty of habitat-templet models lies on the simultaneous consideration of physiological, behavioural, dispersal and demographic parameters in connection with the basic properties of habitats (Korfiatis and Stamou, submitted). Similar to the r-K selection model, habitat templets are graphic models developed on qualitative grounds. The idea of using qualitative models in ecology is ought to Levins (1968). Contrary to quantitative ones accounting mainly for realism and precision, qualitative models account more for realism and generality. Hence, the latter appear rich in biological content and accordingly more informative (Haila 1986), describing more complex phenomena.

According to Korfiatis and Stamou, (submitted), the most important fea-
ture of habitat templets is their capability to describe different mechanisms of
causality than the r-K selection model. In fact, unlike the r-K selection model
in which density dependence is the driving selective force, the principal factor
causing selection in habitat templets relates to elements of the physical environment such as stress, disturbance, favourability and predictability. Habitat
templets have strongly influenced experimental and theoretical work. A pop-
ulation-into-its-environment thinking overwhelms research in the 1990s, and
in the experimental and theoretical work habitats and organisms are viewed
as parts of a feedback-linked system (Korfiatis and Stamou, submitted). For
example, Southwood (1977) stated that the features of an organism interact
with the habitat through an adopted strategy, and organism-environment
relationships are considered as a multivariate-multilevel phenomenon, while
life-history strategies are viewed as a complex of interacting characteristics.
Methodological changes have been considerable. Instead of only demographic parameters, habitat-templet studies involve trade-offs between physiological processes, mechanisms of predator avoidance, ontogenetic processes, such as somatic growth, reproductive tactics, dispersal, diapause etc.
(Southwood 1988).

Despite some advantages, the use of habitat templets is rather limited. To a
great extent this is due to ambiguities concerning the exact definition of concepts as well as to misunderstandings. An example of such difficulties is the
concept of quiescence in oribatids, i.e. the temporary halting of development
below certain temperature thresholds. Indeed, irrespective of whether it is a
recently developed apotypic adaptation in response to actual selective forces
or a realisation of temperature and/or humidity dependent on ancestral
metabolic constraints, quiescence coupled with immediate and left-skewed
metabolic response to increasing temperature is of primordial importance
for Mediterranean arthropods. In fact, it allows animals to overcome hazardous (in autumn and spring) and normally occurring (in winter) low temperatures, while quiescence coupled with rapid response to changing temperature
also enables animals to exploit the slightest temperature increases to accomplish parts of their life cycle development in winter. Apparently, the consideration of various life history attributes within a life history context assigns to
them precise adaptive values, irrespective of their origin. Moreover, several
life history traits may equally be elements of different life history strategies.
For example, Siepel (1994) stated that, although phoresy is normally considered as an r-strategy attribute allowing arthropods the rapid colonisation
of either ephemeral or discretely distributed biotopes, it characterises meso-
stigmatic species showing for the most part K-strategy features.

In the following paragraphs, an attempt to discuss adaptations of Mediterranean arthropods will be undertaken within a habitat templet context as
outlined above. The problem can be stated as follows: as shown in Chapter 7,
specific Mediterranean faunas have not been described and most Mediterra-

nean arthropods are cosmopolitans established prior to the Mediterranean climate. Accordingly, this point concerns the adaptive value of the life history traits of those cosmopolitan species which are able to adjust their life cycle development to elements devising the severity of the Mediterranean habitats.

As shown in Chapter 3, Mediterranean arthropods are characterised by rapid response to changing temperature regimes and low maintenance cost at constant temperatures which, due to low Q_{10} values, increases slightly with increasing temperature. Due to a large temperature plateau, arthropods are able to develop and reproduce over a large range of field temperatures in spring and autumn. In winter Mediterranean arthropods can overcome low temperatures for relatively long periods by entering quiscence and at the same time are able to exploit instantaneous and random increases in temperature to accomplish part of their development. Moreover, metabolic activity of arthropods appears highly susceptible to high temperatures, and consequently high summer temperatures force most arthropods to aestivate, entering an inactive stage. In conclusion, animals appear conservative with respect to the short-term temperature pattern, and the moderately high expenditure at fluctuating temperatures is counterbalanced by low energy transformation capacity at constant temperatures, allowing for survival, further development and reproduction over a large array of temperature conditions. In contrast, summer aestivation and winter quiescence are conformist elements forcing numbers of arthropods to follow the seasonal fluctuations of the Mediterranean climate.

In an attempt to categorise the above characteristics, a classification scheme involving two categories has been proposed. The first category includes conservative elements such as low metabolic activity, low Q_{10} values, large range of tolerance etc. allowing arthropods to survive and reproduce over large ranges of variables. Apparently, elements in the conservative class relate to the animals' endurance under the adverse Mediterranean conditions. The second category includes conformist elements such as rapid response to changing temperatures (accompanied by more efficient transformation of energy), susceptibility to high temperatures, quiescence and aestivation which force demographies to conform with the sesonality of the Mediterranean climate.

In many aspects, most Mediterranean arthropods appear to be demographic conservationists. Indeed, major demographic characteristics such as mortality concentrated in immature instars, low or moderate mortality of adults, slow development, brood protection, relatively long generation time, and low to moderate reproductive values fit the K-selected attributes well. The maintenance of populations showing a low rate of development and low reproductive value involves a long adult life span. Moreover, precocity coupled with limited body size and limited capability for energy storage leads to the dispersal – to some extent – of reproductive effort over time. As was stressed by Norton (1994), long adult life may entail costs invested in conservative ele-

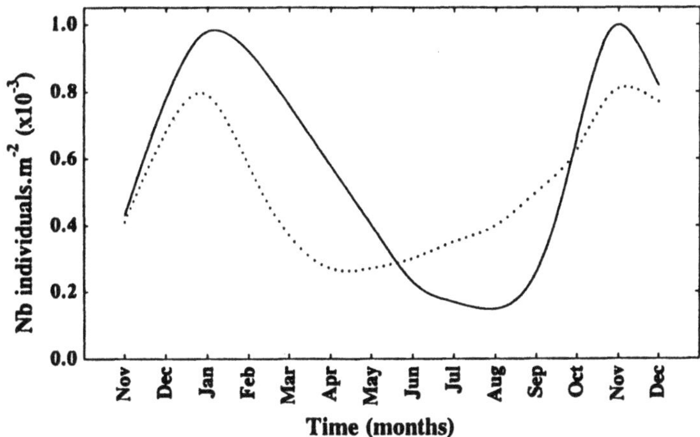

Fig. 8.1. The main phenological types displayed by Mediterranean arthropods. *Solid line* represents right-skewed phenology, *dashed line* left-skewed phenology

ments such as morphological, physiological and behavioural adaptations for survival during extreme environmental conditions, defence and dispersal.

Despite the above conservative traits, additional historical constraints such as voltinism, circumscribed oviposition and physiologically induced winter quiescence and/or summer aestivation force life cycle development and population dynamics to conform with the seasonal oscillations of the Mediterranean climate, inducing generational synchrony. It is noticeable, however, that most conformist characteristics are facultative, offering plasticity to life cycle development. Apparently, conformist characteristics confining egg production and oviposition as well as development of life cycle to favourable periods may offset metabolic constraints improving fertility. Thus, although they tend to be K-strategists, Mediterranean arthropods can hardly be classified on the r-K gradient which appears inadequate to encompass strategies involving both conservative and conformist characteristics. Indeed, as shown in Chapter 6, the majority of population dynamics fall into either the left- or the right-skewed phenologies. As can be realized from Fig. 8.1, in univoltine populations showing left skewed phenologies (e.g. most oribatid species) the maximum rate of demographic events (births and/or deaths) occurs prior to the population density peak in autumn-winter, and population size recovers rapidly after the adverse period in summer. Thus, the phase difference between population density peak and maximum rate of occurrence of demographic events is positive. In contrast, in univoltine populations displaying right-skewed phenologies (e.g. most collembolan species), the maximum rate of demographic events is delayed in relation to the peak of population size, coinciding with the population size decline. In the latter case the phase difference is negative.

The consideration of a phase difference continuum from -6 to 6 months is more appropriate for populations living in regularly fluctuating environments allowing focus on strategies relating to synchronisation with the seasonal environment. In fact, most characteristics of populations showing positive phase difference refer to rapid recovery after adversity, while most characteristics of populations with negative phase difference relate to rapid decline in activity before the onset of the adverse period. In any case, synchronisation capabilities prerequisite the existence of thresholds and narrow selection of crucial demographic parameters and refer to an organism's conformism.

Selection of habitat type is generally conditioned by vectorial environmental factors such as mean, minimum and maximum temperature, relative air humidity, soil properties etc. However, conservative characteristics such as relatively low water budgets are of decisive importance for the dispersal of arthropods within favourable microsites of the subhumid to xeric, but fine-grained Mediterranean habitats. Moreover, the need to avoid a large array of predators as well as possible competitors increasing spatial uncertainty also determines the spatial distribution of Mediterranean arthropods. In contrast, moderate transpiration rates enable them to survive in even temporarily flooded microsites.

As shown in Chapter 7, most Mediterranean arthropods exhibit narrow capabilities for food utilization. Besides, like the majority of worldwide arthropods (e.g. Norton 1994), Mediterranean ones are not able to exploit short-term changes in food resources. However, due to low metabolism they can survive during periods of food shortage. In any case, narrow food selection restricts the spatial distribution of Mediterranean arthropods. Again, conformist water relations such as fluctuations of blood osmolality, integumentary transpiration and differential allocation of water reserves during ontogenesis, as well as the need for behavioural thermoregulation, deposition of eggs and cocoons in favourable microsites and the necessity for taking refuge during summer drought stimulate animals. Thus animals are also forced to conform with the structural adversity of their habitat as well as to undertake the risk of recolonisation of different microsites after man-induced disasters.

Again, strategies driving habitat selection compromise between conservatism and conformism. Conservative demographic features such as sexual reproduction, precocity, dispersal of reproductive effort over time, and density dependence are principal characteristics related to the spatial heterogeneity of habitats. Mainly Allee-type density dependence resulting in aggregative spatial distribution is of primordial importance for understanding of the ecology of Mediterranean arthropods. Indeed, as shown in Chapter 7 Allee-type density dependence accompanied by conformist traits such as narrow food selection, broad weather selection and rapid response to external signals allows arthropods to achieve optimal population size locally.

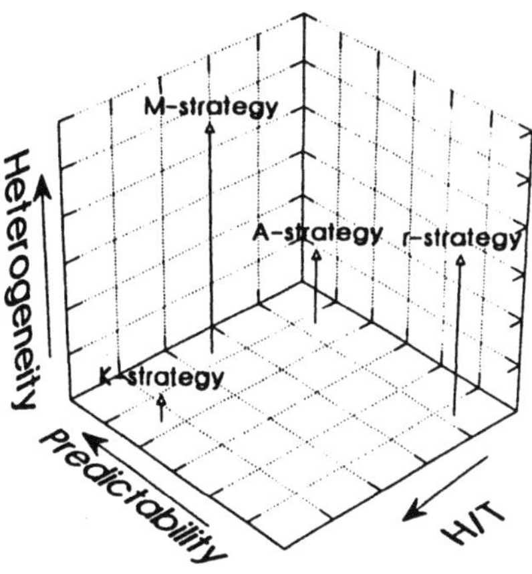

Fig. 8.2. PCA ordination of *r*, *K*, *A*, and *M* (Mediterranean) life history strategies (Asikidis 1989)

For a rough classification of adaptive strategies shown by Mediterranean arthropods, the scheme suggested by Siepel (1994, 1995) can be used. Aiming to tackle applied problems such as the analysis of the effect of management practices and pollution, Siepel (1994) developed ideas concerning the development of a classification key to life history tactics. According to the key proposed in 1995, the characteristics exhibited by most Mediterranean arthropods fit either tactic V (non-phoretic arthropods with obligate diapause or aestivation in their life cycle or synchronisation of their life cycle by quiescence) or tactic XI (non-phoretic arthropods with continuous generation in which development can be slow in some periods, sexual reproduction, and iteroparity) well or even both.

A different classification scheme was developed by Asikidis (1989). To compare life history traits shown by Mediterranean oribatids (considered as M-strategists) with those characterising r-, K- and A-strategists, the author depicted the ordination of life history strategies on the three first axes of a PCA (Fig. 8.2). The first axis represents habitat predictability, the second axis stands for the values of the ratio H/T (H=favourable oviposition time, T=generation time), while the third one relates to the heterogeneity of habitat. Higher predictability, moderate values of the H/T ratio and higher heterogeneity are selective forces driving life history patterns of Mediterranean oribatids.

Even more convenient for classification of the life histories of Mediterranean arthropods is the scheme involving conservatism and conformism. As shown in this section most physiological, behavioural and demographic characters of Mediterranean arthropods fall into either the conservative or the conformist category, and life history strategies imply compromising configurations of conservative and conformist elements. According to Norton (1994), it can be stated that conservatism involving both metabolic compensation and plasticity is resulted from low metabolic rate. In fact, constrained physiology probably entailing low feeding, digestive, metabolic and growth rates is coupled with an incompetence to store energy and mainly with a low response to changing temperature and a large tolerance range. Constrained physiology may drive the development of conservative characteristics such as slow development, relatively long generation time, decreased adult mortality, longevity of adults, and lower investment into reproduction. Thus, moderate maintenance cost and high adult survival in conjunction with density dependent demographies ensure optimal local population size and persistence of Mediterranean arthropods under the constraints of the Mediterranean environment.

Again, constrained physiology involves high food specialisation, rapid response to changing temperature and humidity regimes, minimum and maximum temperature and humidity thresholds, as well as historical constraints such as precocity (which is beneficial to a large clutch size) and relatively small body size, and it relates to arthropods' conformism. Indeed, as shown in this section, synchronising of feeding activity, seasonal shifting in daily activity, sequential activity and developmental patterns, continuous presence of eggs in the abdomen of females (which could be laid opportunistically in response to temporary favourable conditions) combined with circumscribed oviposition and population reserves such as accumulated subsequent instars in winter and/or eggs and early instars in summer result in generational synchrony, forcing population dynamics to conform with the seasonality of the climate. Likewise, water relation, feeding habits and density dependence ensure optimal exploitation of resources. Obviously, low metabolism and vast tolerance ranges, inducing conservatism, coupled with environmental thresholds, rapid response to changing temperature and humidity regimes, and narrow food utilization, inducing conformism, are favoured forces driving life history strategies of Mediterranean arthropods. As has been shown, adaptive emphasis is put for the most part on low maintenance cost, efficient exploitation of resources, adult longevity, and low reproductive mortality counterbalanced by low viability of the immature stages.

References

Abott I (1984) Changes in the abundance and activity of certain soil and litter fauna in the Jarrah forest of Western Australia after a moderate intensity fire. Aust J Soil Res 22: 463-469

Argyropoulou MD (1993) Dynamics and activity of collembolan in the soil subsystem of an evergreen-sclerophyllous formation in Mt. Hortiatis. PhD Thesis, University of Thessaloniki (in Greek with English summary)

Argyropoulou MD, Stamou GP (1993) Respiratory activity of the collembolan Onychiurus meridiatus. J Insect Physiol 39: 217-222

Argyropoulou MD, Asikidis MD, Iatrou GD, Stamou GP (1993) Colonisation patterns of decomposing litter in a maquis ecosystem. Eur J Soil Biol 29: 183- -191

Argyropoulou MD, Stamou GP, Iatrou GD (1994) Temporal and spatial distribution patterns of Collembola in a patchy environment. Eur J Soil Biol 30: 63-69

Arianoutsou-Faraggitaki M (1979) Biological activity after fire in a phryganic ecosystem. PhD Thesis, University of Thessaloniki (in Greek with English summary)

Aschmann H (1973) Distribution and peculiarity of Mediterranean ecosystems. In: Di Castri F, Mooney HA (eds) Mediterranean-type ecosystems: origin and structure. Springer, Berlin Heidelberg New York, pp 11-19

Asikidis MD (1989) Population dynamics and activity of oribatid mites from an evergreen-sclerophyllous formation (Hortiatis). PhD Thesis, University of Thessaloniki (in Greek with English summary)

Asikidis MD, Stamou GP (1991) Spatial and temporal patterns of oribatid mite community in an evergreen-sclerophyllous formation (Hortiatis, Greece). Pedobiologia 35: 53-63

Asikidis MD, Stamou GP (1992) Phenological patterns of oribatid mites in an evergreen-sclerophyllous formation (Hortiatis, Greece). Pedobiologia 36: 359--372

Athias-Binche F (1981) Écologie des uropodides édaphiques (Arachnides: Parasitoformes) de trois écosystèmes forestiers. 1. Introduction, matériel, biologie. Vie Milieu 31: 137-147

Athias-Binche F (1984) La phoresie chez les acariens uropodides (Anactinotriches), une stratégie écologiques originale. Acta Oecol/ Oecol Gen 5: 119-133

Athias-Binche F (1985) Analyse démographiques des population d' Uropodides (Arachnides: Anactinotriches) de la hêtraie de la Massane, France. Pedobiologia 28: 225-253

Athias-Binche F (1987) Modalités de la cicatrisation des écosystèmes méditerranéens après incendie: cas de certains arthropods du sol. 3. Les acariens uropodides. Vie Milieu 37: 39-52

Athias-Binche F (1995) Is the initiation of dispersal genetically fixed? The case of phoretic mites. In: Kropczynska D, Boczek J, Tomczyk A (eds) The Acari Dabor, Warszawa. pp 149-162

Baker GH (1978a) The post-embryonic development and life history of the millipede, Ommatoiulus moreletii (Diplopoda: Iulidae), introduced in south-eastern Australia. J Zool Lond 186: 209-228.

Baker GH (1978b) The population dynamics of the millipede, Ommatoiulus moreletii (Diplopoda: Iulidae). J Zool Lond 186: 229-242.

Baker GH (1979a) The activity patterns of Ommatoiulus moreletii (Diplopoda: Iulidae) in South Australia. J Zool Lond 188: 173-183.

Baker GH (1979b) Eruptions of the introduced millipede, Ommatoiulus moreletii (Diplopoda, Iulidae), in Australia, with notes on the native Australiosoma castaneum (Idiplopoda, Paradoxosomatidae). S Aust Nat 53: 36-41

Baker GH (1980) The water and temperature relationships of Ommatoiulus moreletii (Diplopoda: Iulidae). J Zool Lond 190: 97-108

Baker GH (1984) Distribution, morphology and life history of the millipede Ommatoiulus moreletii (Diplopoda: Iulidae) in Portugal and comparison with Australian populations. Aust J Zool 32: 811-822

Baker GH (1985) The distribution of the Portuguese millipede Ommatoiulus moreletii (Diplopoda:Iulidae) in Australia. Aust J Ecol 10: 249-259

Barra JA, Belgnaoui S, Poinsot-Balaguer N (1989) Sécheresse, froid même stratégies, un example type: le collembole Folsomides angularis. Bull Ecol 20: 65-66

Belgnaoui S, Barra JA (1988) Cytochrome oxidase activity in the anhydrobiotic Collembola Folsomides angularis (Insecta, Apterygota). Pedobiologia 32: 283--291

Belgnaoui S, Barra JA (1989) Water loss and survival in the anhydrobiotic Collembola Folsomides angularis (Insecta). Rev Ecol Biol Sol 26: 123-132

Bercovitz K, Warburg MR (1985) Developmental patterns in two populations of the millipede Archispirostreptus syriacus (De Saussure) in Israel. Bijd Dierkd 55: 37-46

Bercovitz K, Warburg MR (1988) Factors affecting egg-laying and clutch size of Archispirostreptus tumuliporus judaicus (Attems) (Myriapoda, Diplopoda) in Israel. Soil Biol Biochem 20: 869-874

Bernini F (1984) Main trends of oribatid mite biogeography in the central-west Mediterranean. In: Griffiths DA, Bowman CE (eds) Acarology VI, vol 2. Hellis Horwood Publishers, 932-940

Berthet P (1964) L' activité des oribates d' une chênaie. Mem Inst Rech Sci Nat Bel 152: 1-152

Bertrand M, Janati-Idrissi A, Lumaret J-P (1987) Étude expérimentale des facteurs de variation de la consommation de la litière de Quercus ilex L. et Q. pubenscens Wild. par Glomeris marginata. (V.) (Diplopoda: Glomeridae). Rev Ecol Biol Sol 24: 359-368

Bigot L, Bodot P (1973) Contribution à l'étude biocenotique de la garrique à Quercus coccifera III. Dynamique de la zoocenose d' invertébrés. Vie Milieu 23: 251-267

Block W (1977) Oxygen consumption of the terrestrial mite Alaskozetes antarcticus (Acari: Cryptostigmata). J Exp Biol 68: 69-87

Block W (1979) Oxygen consumption of the Antarctic springtail Parisotoma octooculata (Willem) (Isotomidae). Rev Ecol Biol Sol 16: 227-233

Block W (1985) Arthropod interactions in an antarctic terrestrial community In: Siegfried WR, Condy PR, Laws RM (eds) Antarctic nutrient cycles and food webs. Springer, Berlin Heidelberg New York, pp 614-619

Block W (1996) Cold or drought - the lesser of two evils for terrestrial arthropods? Eur J Entomol 93: 325-339

Block W, Tilbrook PJ (1977) Effect of long term storage on the oxygen uptake of Cryptopygus antarcticus (Collembola). Oikos 29: 284-289

Block W, Young SR (1978) Metabolic adaptations of antarctic terrestrial microarthropods. Comp Biochem Physiol 61A: 363-368

Block W, Harrison PM, Vannier G (1990) A comparative study of patterns of water loss from two antarctic springtails (Insecta, Collembola). J Insect Physiol 36: 181-187.

Blower JG (1969) Age structure of millipede populations in relation to activity and dispersion. Publ Syst Assa 8: 209-216

Bond W (1983) On alpha diversity and the richness of the Cape flora: a study in southern Cape fynbos. In: Kruger FJ, Mitchell DT, Jarvis JVM (eds) Mediterranean-type ecosystems: the role of nutrients. Springer, Berlin Heidelberg New York, pp 337-356

Byzova JB (1967) Respiration metabolism in some millipedes. Rev Ecol Biol Sol 4: 611-624

Calow P (1978) Ecology, evolution and energetics: a study in metabolic adaptation. Adv Ecol Res 10: 1-62

Cancela da Fonseca JP, Hadjibiros K (1977) Le modele matriciel determinist de Leslie et ses application en dynamique des populations. Acta Biotheretica 26: 239-261

Cassagnau P (1986) Les écomorphose des collemboles: 1. Déviations de la morphogenèse et perturbations histophysiologiques. Ann Soc Entomol Fr (NS) 22: 7-33

Charnov EL, Shaffer W (1973) Life history consequences of natural selection: Cole's result revised. Am Nat 107: 791-793

Cloudsley-Thompson JL (1951) On the responses to environmental stimuli and the sensory physiology of millipedes (Diplopoda). Proc Zool Soc Lond 121: 253-277

Cloudsley-Thompson JL (1953) The significance of fluctuating temperatures on the physiology and ecology of insects. Entomologist 86: 183-189

Cloudsley-Thompson JL (1965) On the function of the sub-elytral cavity in desert Tenebrionidae (Col.). Entomol Mon Mag 100: 148-151

Cloudsley-Thompson JL (1969) Acclimation, water and temperature relations of the wood lice Metoponothrus pruinosus and Periscyphis jannonei in the Sudan. J Zool Lond 158: 267-276

Cloudsley-Thompson JL (1975) Adaptations of Arthropoda to arid environments. Annu Rev Entomol 20: 261-283

Cloudsley-Thompson JL (1981) A comparison of rhythmic locomotory activity in tropical forest Arthropoda with that in desert species. J Arid Environ 4: 327-334

Cloudsley-Thompson JL (1982) Desert adaptations in spiders. Sci Rev Arid Zone Res 1: 1-14

Cloudsley-Thompson JL (1983) Desert adaptations in spiders. J Arid Environ 6: 307-317

Cloudsley-Thompson JL (1988) Evolution and adaptation of terrestrial arthropods. Springer, Berlin Heidelberg New York, 141 pp

Cloudsley-Thompson JL, Constantinou C (1983) Transpiration from forest dwelling and woodland Mygalomorphae (Araneae). Int J Biometeorol 27: 69- -74

Cloudsley-Thompson JL, Crawford CS (1970) Water and temperature relations, and diurnal rhythms of scolopedromorph centipedes. Entomol Exp & Appl 13: 187-193

Cole LM (1954) The population consequences of life history phenomena. Q Rev Biol 29: 103-137

Crawford CS, Warburg MR (1982) Water balance and apparent oocyte resorption in desert millipedes. J Exp Zool 222: 215-226

Crawford CS, Goldberg S, Warburg MR (1986) Seasonal water balance in Archispirostreptus syriacus (Diplopoda: Sprirostreptidae) from mesic and xeric Mediterranean environments. J Arid Environ 10: 127-136

Crawford CS, Bergovitz K, Warburg MR (1987) Regional environments, life history patterns, and habitat use of spirostreptid millipedes in arid regions. Zool J Linn Soc 89: 63-88

Daget P (1984) Introduction à une théorie générale de la Méditerranéité. Bull Soc Bot Fr 131: 31-36

David JF (1982) Variabilité dans l'espace et dans le temps des cycles de vie des deux populations de Cylidroiulus nitidus (Verhoeff) (Julida). Rev Ecol Biol Sol 19: 411-425

David JF (1987) Consommation annuelle d'une litière de chêne par une population adulte du diplopode Cylindroiulus nitidus. Pedobiologia 30: 299- -310

David JF (1995) Seasonal abundance of millipedes in a Mediterranean oak forest (Southern France). Isr J Zool 41: 23-31

De Izzara DC (1977) Les effets de l'amploi du feu sur les microarthropodes du sol dans la région semi-aride pampeenne. In: Soil organisms as components of ecosystems. Ecol Bull 25: 357-365

Diamantopoulos J, Pirintsos SA, Margaris NS, Stamou GP (1994) Variation in Greek phrygana vegetation in relation to soil and climate. J Veg Sci 5: 355-360

Di Castri F (1973) Climatological comparisons between Chile and western coast of N. America. In: Di Castri F, Mooney HA (eds) Mediterranean-type Ecosystems. Springer, Berlin Heidelberg New York, pp 21-36

Di Castri F, Vitali-Di Castri V (1981) Soil fauna of Mediterranean - climate regions. In: Di Castri F, Goodal DW, Specht RL (eds) Ecosystems of the world. II. Mediterranean-type shrublands. Elsevier, Amsterdam, pp 445-478

Dwarakanath SK (1971a) The influence of body size and temperature upon the oxygen consumption in the millipede, Spirostreptus asthenes (Pocock). Comp Biochem Physiol 38: 351-358

Dwarakanath SK (1971b) Effect of temperature on the oxygen consumption in the tropical millipede Harpurostreptus sp. Proc Indian Biol Sci 68: 4-7

Dwarakanath SK, Berlin OGW, Pantian RW (1973) Oxygen consumption as a function of size and temperature in the pill millipede Arthrosphaera disticta (Pocock). Monit Zool Ital 7: 43-50.

Economidou E (1976) La répartition des phrygana en Grèce et ses rapports avec le climat et l' influence anthropogène. Doc Phytosociol 15: 45-56.

Edmonds SJ, Specht MM (1981) Dark island heathland, South Australia: faunal rhythms. In: Specht RL (ed) Heathlands and related shrublands of the world. B. Analytical studies. Elsevier, Amsterdam, pp 15-27

Edney EB (1977) Water balance in land arthropods. Springer, Berlin Heidelber New York, 286 pp

Fouseki E (1979) Decomposition and soil metabolism in a phryganic ecosystem. PhD Thesis, University of Thessaloniki (in Greek with English summary)

Fouseki E, Margaris NS (1981) Soil metabolism and decomposition in a phryganic (eastern Mediterranean) ecosystem. Oecologia 50: 417-421

Fox BJ, Fox MD (1986) Resilience of animal and plant communities to human disturbance. In: Hopkins AJM, Lamont BB (eds) Resilience in Mediterraneantype ecosystems. Dr W Junk , The Hague, pp 39-64

Fujita H (1954) An interpretation of the change in type of the population density effect upon the oviposition rate. Ecology 35: 253-257

Gadgil M, Bossert W (1970) Life history consequences of natural selection. Am Nat 104: 1-24

Ghabbour SI (1983) Population density and biomass of terrestrial isopods (oniscoids) in the xero-Mediterranean agro-ecosystems of Mariut region, Egypt. Ecol Mediterr 9: 3-16

Ghabbour SI, Rizk MA (1979) Patterns of growth and body water content in xeric and mesic isopods. Pedobiologia 19: 18-25

Giliomee JH (1989) First record of western flower thrips, Frankliniella occidentalis (Pergande) (Thysanoptera: Thripidae, from South Africa. J Entomol Soc South Afr 8: 179-182

Giliomee JH (1992) Factors disturbing insects in a fynbos reserve. Proc VI Int Conf Mediterranean Climate Ecosystems (MEDECOS), Meleme (Crete) 1991, pp 133-139

Greenslade PJM (1972a) Distribution patterns of Priochirus (Coleoptera: Staphylinidae) in the Solomon Islands. Evolution 26: 130-142

Greenslade PJM (1972b) Evolution in the staphylinid genus Priochirus (Coleoptera). Evolution 26: 203-220

Greenslade P (1982) Origin of the Collembola fauna of arid Australia. In: Barker R, Greenslade PJM (eds) Evolution of the flora and fauna of arid Australia. Peacock Publications, Adelaide, pp 267-272

Greenslade P (1983) Adversity selection and the habitat templet. Am Nat 122: 352-365

Grime JP (1977) Evidence for the existence of three primary strategies in plants and its relevance to ecological evolutionary theory. Am Nat 111: 1169-1194

Grime JP (1979) Plant strategies and vegetation processes. Wiley, New York

Gromysz-Kalkowska K (1970) The influence of body weight, external temperature, seasons of the year and fasting on respiratory metabolism in Polydesmus complanatus L. (Diplopoda). Folia Biol 18: 311-326

Gromysz-Kalkowska K (1974) The effect of some exogenous factors and body weight on the oxygen consumption in Glomeris connexa (C.L. Koch) (Diplopoda). Folia Biol 22: 37-49

Gromysz-Kalkowska K, Tracz H (1983) The effect of temperature, food kind and body weight on the oxygen consumption of Proteroiulus fuscus. Ann Warsaw Agric Univ - SGGW-AR Wood Technol 30: 35-42

Hadley NF (1994) Water relations of terrestrial arthropods. Academic Press, London, 356 pp

Hadley NF, Quinlan MC (1984) Cuticular transpiration in the isopod Porcellio laevis: chemical and morphological factors involved in its control. Symp Zool Soc Lond 53: 97-107

Haila Y (1986) On the semiotic dimension of ecological theory: the case of island biogeography. Biol Phil 1: 377-388

Hanski I (1982) Dynamics of regional distribution: the core and satellite species hypothesis. Oikos 38: 210-221

Hanski I (1991) Single species metapopulation dynamics: concepts, models and observations. Biol J Linn Soc 42: 17-38

Hanski I, Gilpin M (1991) Metapopulation dynamics: brief history and conceptual domain. Biol J Linn Soc 42: 3-16

Heath J, Bocock KL, Mountford MD (1974) The life history of the millipede Glomeris marginata (Villers) in north-west England. Symp Zool Soc Lond 32: 433-462

Holling CS (1959) Some characteristics of simple types of predation and parasitism. Can Entomol 91:385-398

Hopkin SP, Read HJ (1992) The biology of millipedes. Oxford Sciences Publications, Oxford New York Tokyo, 233 pp

Hornung E, Warburg MR (1996) Intra-habitat distribution of terrestrial isopods. Eur J Soil Biol 32:179-185

Horvart I, Glavac V, Ellenberg H (1974) Vegetation Südosteuropas (Vegetation of Southeast Europe) Gustav Fischer, Stuttgart

Housard C, Escarre J, Romane F (1980) Development of species diversity in some Mediterranean plant communities. Vegetatio 43: 59-72

Humphreys WF (1979) Production and respiration in animal populations. J Anim Ecol 48: 27-253

Iatrou GD (1989) Dynamics and activity of the diplopod Glomeris balcanica in the soil subsystem of an evergreen-sclerophyllous formation in Mt. Hortiatis. PhD Thesis, University of Thessaloniki (in Greek with English summary)

Iatrou GD, Stamou GP (1989a) Preliminary studies on certain macroarthropod groups of a Quercus coccifera formation (Mediterranean-type ecosystem) with reference to the diplopod Glomeris balcanica. Pedobiologia 33: 301-306

Iatrou GD, Stamou GP (1989b) Seasonal activity patterns of Glomeris balcanica (Diplopoda: Glomeridae) in an evergreen-sclerophyllous formation in northern Greece. Rev Ecol Biol Sol 26: 491-503

Iatrou GD, Stamou GP (1990) Studies on the life cycle of Glomeris balcanica (Diplopoda, Glomeridae) under laboratory conditions. Pedobiologia 34: 173- -181

Iatrou GD, Stamou GP (1991) The life cycle and spatial distribution of Glomeris balcanica (Diplopoda, Glomeridae) in an evergreen-sclerophyllous formation in northern Greece. Pedobiologia 35: 1-10

Johnson SD, Bond WJ (1992) Convergent floral evolution in a guild of butterfly pollinated fynbos plants. Proc VI Int Conf Mediterranean Climate Ecosystems (MEDECOS), Meleme (Crete) 1991, pp 228-233

Joose ENG (1971) Ecological aspects of aggregation in Collembola. Rev Ecol Biol Sol 8: 91-97

Juchault P, Rigaud T, Mocquard JP (1993) Evolution of sex determination and sex ratio variability in wild populations of Armadillidium vulgare (Latr.) (Crustacea, Isopoda): a case study in conflict resolution. Acta Oecologica 14: 547-562

Karamaouna MA (1990) On the ecology of the soil macroarthropod community of a Mediterranean pine forest (Sophico, Peloponnese, Greece) Bull Ecol 21: 33-42

Karamaouna M, Geoffroy J-J (1985) Millipedes of a maquis ecosystem (Naxos island, Greece): preliminary description of the population (Diplopoda). Bijd Dierkd 55: 113-115

Kheirallah AM (1980) Aspects of distribution and community structure of isopods in the Mediterranean coastal desert of Egypt. J Arid Environ 3: 69-74

Kirk AA, Ridsdill-Smith TJ (1986) Dung beetle distribution pattern in the Iberian peninsula. Entomophaga 31: 183-190

Korfiatis KJ, Stamou GP (1994) Emergence of new fields in ecology: the case of life history studies. Hist Philos Life Sci 16: 97-116

Korfiatis KJ, Stamou GP (1997) Habitat templets and changing world view of ecology. (submitted)

Lamont BB (1994) Triangular trophic relationships in Mediterranean-climate Western Australia. In: Arianoutsou M, Groves RH (eds) Plant-animal interactions in Mediterranean-type ecosystems. Kluwer, Dordrecht, pp 83-89

Lamotte M, Bladin P (1989) Originalité et diversité des écosystèmes méditerranéens terrestres. Biol Gallo-Hell 16: 5-36

Lebrun P (1971) Écologie et biocénotique de quelques peuplements d' arthropodes édaphiques. Mem Inst Rech r Sci Nat Belg 165: 1-123.

Lebrun P, Ruymbecke M (1971) Intêrét écologique de la relation entre la température et la durée du dévelopement des oribates. Acarologia 13: 176-185

Levins R (1968) Evolution in changing environments. Princeton University Press, Princeton

Logan JA, Wollkind DJ, Hoyt S, Tanigoshi LK (1976) An analytic model for description of temperature dependent rate phenomena in arthropods. Environ Entomol 5: 1133-1140.

Lopes CM, da Gama MM (1994) The effect of fire on collembolan populations of Mata da Maga-
    raça (Portugal). Eur J Soil Biol 30: 133-141
Lossaint P, Rapp M (1971) Répartition de la matière organique, productivité et cycles des élé-
    ments minéraux dans des écosystèmes de climat méditerranéen. In: Duvigneaud P (ed)
    Production des écosystèmes forestiers. Ecologie et conservation, vol 4. Unesco, Paris, pp
    597-617
Luxton M (1972) Studies on the oribatid mites of a Danish beech wood soil. Pedobiologia 15:
    161-200
Luxton M (1975) Studies on the oribatid mites of a Danish beech wood soil II. Biomass, calori-
    metry, and respirometry. Pedobiologia 15: 161-200
Luxton M (1981a) Studies on the oribatid mites of a Danish beech wood soil. IV. Developmental
    biology. Pedobiologia 21: 312-340
Luxton M (1981b) Studies on the oribatid mites of a Danish beech wood soil. VI. Seasonal popu-
    lation changes. Pedobiologia 21: 387-409
MacArthur RH, Wilson EO (1967) The theory of island biogeography. Princeton University
    Press, Princeton
Magioris SN (1985) Preliminary data on the annual evolution and activity of the soil mesofauna
    in two insular ecosystems. Rapp Comm int Mer Mediterr 29: 117-121
Magioris SN, Tsiourlis GM (1992) The role of the shrubs Quercus coccifera in the community
    structure of soil arthropods in two insular ecosystems (Is. of Naxos,Cyclades,Greece). Proc VI
    Int Conf Mediterranean Climate Ecosystems (MEDECOS), Meleme (Crete) 1991, pp 140-146
Maggs J, Pearson CJ (1977a) Litter fall and litter layer decay in coastal scrub at Sydney, Austra-
    lia. Oecologia 31: 227-237
Maggs J, Pearson CJ (1977b) Minerals and dry matter in coastal scrub and grassland at Sydney,
    Australia. Oecologia 31: 227-237.
Majer JD (1984) Short-term responses of soil and litter invertebrates to a cool autumn burn in
    Jarrah (Eucalyptus marginata) forest in Western Australia. Pedobiologia 26: 229-247
Mardiris TA (1992) The effect of reforestation on the structure of the Mediterranean-type eco-
    systems. PhD Thesis, Aegean University, Mytilene (in Greek with English summary)
Margaris NS (1976) Structure and dynamics in a phryganic (East Mediterranean) ecosystem.
    J Biometeorol 3: 249-259
Margaris NS (1981) Adaptive strategies in plants dominated Mediterranean type ecosystems.
    In: Di Castri F, Goodal W, Specht RL (eds) Mediterranean type shrublands. Elsevier, Amster-
    dam, pp 309-315
Mason CF (1970) Food, feeding rates and assimilation in woodland snails. Oecologia 4: 358-370
Matsakis J, Tsiourlis GM, Karamaouna M, Legakis A, Paraschi L, Blandin P, Lamotte M (1992)
    Étude d' un écosystème de maquis à Juniperus phoenica L. (Naxos, Cyclades, Grèce): pré-
    sentation générale. Bull Ecol 23: 49-58
Matthiessen JN (1991) Population phenology of white-fringed weevil, Graphognathus leuco-
    loma (Coleoptera: Curculionidae), in a pasture in a Mediterranean-climate region of Aus-
    tralia. Bull Entomol Res 81: 283-289
Matthiessen JN, Ridsdill Smith TJ (1991) Populations of African black beetle, Heteronychus ara-
    tor (Coleoptera: Scarabaeida), in a Mediterranean climate region of Australia. Bull Entomol
    Res 81: 85-91
Maurer BA (1990) The relationship between distribution and abundance in a patchy environ-
    ment. Oikos 58: 181-189
McQueen DJ, Steel CGH (1980) The role of photoperiod and temperature in the initiation of
    reproduction in the terrestrial isopod Oniscus asellus Linnaeus. Can J Zool 58: 235-240
Mena J, Galante E, Lumbreras CJ (1989) Daily flight activity of Scarabaeidae and Geotrupidae
    (Col.) and analysis of the factors determining this activity. Ecol Mediterr 15: 69-80
Metz LJ (1971) Vertical movement of acarina under moisture gradients. Pedobiologia 11: 262-
    268
Mitchell DT, Coley PGF, Webb S, Allsop N (1986) Litter fall and decomposition process in the
    coastal fynbos vegetation, S. W. Cape, S. Africa. J Ecol 74: 977--993
Mitchell M (1979) Energetics of oribatid mites (Acari: Cryptostigmata) in an aspen woodland
    soil. Pedobiologia 19: 89-98

Mitrakos K (1980) A theory for mediterranean plant life. Acta Oecologica, Oecol Plant 1: 245-252

Mooney HA (1977) Southern coastal scrub. In: Barbour M, Major J (eds) Terrestrial vegetation of California. Wiley, New York, pp 180-191

Nahal I (1981) The Mediterranean climate from a biological view point. In: Di Castri F, Goodal DW, Specht RL (eds) Ecosystems of the world. II. Mediterranean-type shrublands. Elsevier, Amsterdam, pp 63-86

Naveh Z, Whittaker RH (1979) Structural and floristic diversity of shrublands and woodlands in northern Israel and other Mediterranean areas. Vegetatio 41: 171-190

Newell RC, Wieser W, Pye VI (1974) Factors affecting oxygen consumption in the wood louse Porcellio scaber. Oecologia 16: 31-51

Nilsen ET, Muller WH (1981) Phenology of the drought deciduous scrub Lotus scoparius: climatic controls and adaptive significance. Ecol Monogr 51: 307- -322

Norton RA (1994) Evolutionary aspects of oribatid mite life histories and consequences for the origin of the Astigmata. In: Houck M (ed) Mites. Ecological and evolutionary analyses of life-history patterns. Chapman and Hall, New York, pp 99-135

Norton RA, Palmer SC (1991) The distribution, mechanisms and evolutionary significance of parthenogenesis in oribatid mites. In: Schuster R, Murphy PW, (eds) Reproduction, Development and life-history strategies. Chapman and Hall, London, pp 108-136

Pantis JD (1987) Structure, dynamics and management of the asphodele deserts in Thessaly. PhD Thesis, University of Thessaloniki (in Greek with English summary)

Pantis JD, Mardiris TA (1992) The effects of grazing and fire on degradation processes of Mediterranean ecosystems. Isr J Bot 41: 233-242

Pantis JD, Stamou GP, Sgardelis SS (1988) Activity patterns of surface ground fauna in asphodel deserts (Thessalia, Greece). Pedobiologia 32: 81-87

Papatheodorou E (1996) The effect of grazing on the structure and dynamics of vegetation and on the dynamics of nutrients of an evergreen-sclerophyllous formation in Mt. Hortiatis. PhD Thesis, University of Thessaloniki (in Greek with English summary)

Papatheodorou E, Pantis JD, Stamou GP (1993) The effects of grazing on growth, spatial pattern and age structure of Quercus coccifera. Acta Oecologia 14: 589-602

Paraschi L (1988) Study of spiders in Mediterranean maquis ecosystems of southern Greece. PhD Thesis, University of Athens (in Greek)

Paraskevopoulos SP, Iatrou GD, Pantis JD (1994) Plant growth strategies in evergreen-sclerophyllous shrublands (maquis) in central Greece. Vegetatio 115: 109-114

Paris OH (1963) The ecology of Armadillidium vulgare (Isopoda: Oniscoidea) in California grassland: food, enemies and weather. Ecol Monogr 33: 1-22.

Penteado CHS, Mendes EG (1977) Respiratory metabolism and tolerance in a tropical millipede, Rhinocricus padbergi. I. The structure of the tracheal pocket and the respiratory rate at normoxic conditions and 25 0C. Rev Brasil Biol 37: 431-446

Petanidou T (1991) Pollination ecology in a phryganic ecosystem. PhD Thesis, Aristotle University, Thessaloniki, Greece (in Greek with English summary)

Petanidou T, Vokou D (1990) Pollination and pollen energetics in Mediterranean ecosystems. Am J Bot 77: 986-992

Petanidou T, Vokou D (1993) Pollination ecology of Labiatae in a phryganic (eastern Mediterranean) ecosystem. Am J Bot 80: 892-899

Peters RH (1983) The ecological implications of body size. Cambridge University Press, Cambridge

Petersen H (1981) Open gradient diver respirometry modified for terrestrial microarthropods. Oikos 37: 265-272

Petersen H, Luxton MS (1982) A comparative analysis of soil fauna populations and their role in decomposition processes. 4. Structure and size of soil animal populations. Oikos 39: 288-388

Pianka E (1970) On rand K-selection. Am Nat 106: 581-588

Poinsot N (1968) Cas d'anhydrobiose chez le collembole Subisotoma variabilis Gisin. Rev Ecol Biol Sol 4: 585-586

Poinsot-Balaguer N (1988) Stratégies adaptative des arthropodes du sol en région méditerranéenne. In: Di Castri F, Floret C, Rambal S, Roy J (eds) Time scales and water stress. Proc 5th Int Conf on Mediterranean Ecosystems. IUBS, Paris, pp 511-539

Poinsot-Balaguer N (1990) Des insectes résistants à la sécheresse. Sécheresse 4: 265-271

Poinsot-Balaguer N, Barra JA (1991) L'anhydrobiose: un problème nouveau chez collemboles (Insecta). Rev Ecol Biol Sol 28: 197-205

Poinsot-Balaguer N, Sadaka N (1986) Distribution saisonnière et vertical d' une population d' Onychiurus zschokkei Handschin (collembole) dans une litière d' une forêt de chêne vert (Quercus ilex Linnè) de la région méditerranéenne française. Ecol Mediterr 12: 9-13

Pomeroy DE, Rwakaikara D (1975) Soil arthropods in relation to grassland burning. East Afr Agric For J 41: 114-118

Postle AC (1985) Density and seasonality of soil and litter invertebrates at Dwellingup. In: Greenslade P, Majer JD (eds) Soil and litter invertebrates of Australian Mediterranean-type ecosystems. WAIT School Biol Bull 12: 18-20

Postle AC, Majer JD, Bell DT (1991) A survey of selected soil and litter invertebrate species from the northern jarrah (Eucalyptus marginata) forest of Western Australia, with particular reference to soil-type, stratum, seasonality and the conservation of forest fauna. In: Lunney D (ed) Conservation of Australian forest fauna. Royal Zoological Society of NSW, Mosman, pp 193-   -203

Precht H (1958) Concepts of the temperature adaptation of unchanging reaction of cold-blooded animals. In: Prosser CL (ed) Physiological adaptation. American Physiological Society, Bethesda, pp 50-78

Prosser CL (1973) Comparative animal physiology, 3rd edn. Saunders, Philadelphia

Quezel P, Barbero M (1982) Definition and characterization of Mediterraneantype ecosystems. In: Quezel P (ed) Définition et localisation des ecosystèmes mediterraneens terrestres. Ecol Mediterr 8: 15-29

Quinlan MC, Hadley NF (1983) Water relations of the terrestrial isopods Porcellio laevis and Porcellionides pruinosus (Crustacea, Oniscoidea). J Comp Physiol 151: 155-161

Radea K (1989) A study of litter, decomposition and the community of arthropods in an insular ecosystem of Pinus halepensis. PhD Thesis, University of Athens (in Greek with English summary)

Raus T (1979a) Die Vegetation Ostthessaliens (Griechenland). I. Vegetationszonen und Höhenstufen. Bot Jahrb Syst 100: 564-601

Raus T (1979b) Die Vegetation Ostthessaliens (Griechenland). II. Querceteaillicis und Cisto-Micromerietea. Bot Jahrb Syst 101: 17-82

Read DJ, Mitchell DT (1983) Decomposition and mineralization processes in Mediterranean-type ecosystems and heathlands of similar structure. In: Kruger FG, Mitchell DT, Jarvis JUM (eds) Mediterranean-type ecosystems: the role of nutrients. Springer, Berlin Heidelberg New York, pp 208-232

Riddle WA (1978) Respiratory physiology of the desert grassland scorpion Paruroctonus utahensis. J Arid Environ 1: 243-251

Ridsdill-Smith TJ (1986) The effect of seasonal changes in cattle dung on egg production by two species of dung beetles (Coleoptera: Scarabeidae) in southwestern Australia. Bull Entomol Res 76: 63-68

Saulnier l, Athias-Binche F (1986) Modalités de la cicatrisation d'écosystèmes méditerranéens après incendie: cas de certains arthropods du sol. 2. Les myriapodes édaphiques. Vie Milieu 36: 191-204

Sauvage C (1961) Recherches géobotaniques sur la chêne liège au Maroc. Trav Inst Sci Cherifien, Sér Bot 21: 402

Sgardelis SP (1988) The effect of fire on the consumers of a phryganic ecosystem]. PhD Thesis, University of Thessaloniki (in Greek with English summary)

Sgardelis SP (1995) Body size, sex ratio and fecundity of soil Cryptostigmata along an altitudinal gradient. Acta Zool Fenn 196: 258-259

Sgardelis S, Margaris NS (1983) Seasonal activity of soil fauna in a phryganic (eastern Mediterranean) ecosystem. In: Margaris NS, Arianoutsou-Faraggitaki M, Reiter R (eds) Adaptations to terrestrial environments. Plenum Press, New York, pp 31-43

Sgardelis SP, Margaris NS (1993) Effects of fire on soil microarthropods of a phryganic ecosystem. Pedobiologia 37: 83-94

Sgardelis S, Stamou G, Margaris NS (1981) Structure and spatial distribution of soil arthropods in a phryganic (eastern Mediterranean) ecosystem. Rev Ecol Biol Sol 18: 221-230.

Sgardelis SP, Sarkar S, Asikidis MD, Cancela da Fonseca JP, Stamou GP (1993) Phenological patterns of soil microarthropods from three climate regions. Eur J Soil Biol 29: 49-57

Sgardelis SP, Pantis JD, Argyropoulou MD, Stamou GP (1995) Effects of fire on soil macroinvertebrates in a Mediterranean phryganic ecosystem. Int J Wildl Fire 5:113-121

Sheppard AW, Michalakis DT, Briese DT, Thomann T (1992) Spatial scale int the interaction between seed-head weevils and their host plants (Asteraceae, Cardueae), in Mediterranean sheep-grazing system. Proc VI Int Conf Mediterranean Climate Ecosystems (MEDECOS), Meleme (Crete) 1991, pp 183-190

Siepel H (1994) Life-history tactics of soil microarthropods. Biol Fertil Soils 18: 263-278

Siepel H (1995) Applications of microarthropod life-history tactics in nature management and ecotoxicology. Biol Fertil Soils 19: 75-83

Snider RM (1981) The reproductive biology of Polydesmus inconstans (Diplopoda: Polydesmidae) at constant temperatures. Pedobiologia 22: 354-365

Sømme L (1995) Invertrebrates in Hot and Cold Environments. Springer, Berlin, 275 pp

Sømme L (1996) Anhydrobiosis and cold tolerance in tardigrades. Eur J Entomol 93: 349-357

Southwood TRE (1977) Habitat, the templet for ecological strategies. J Anim Ecol 46: 337-365

Southwood TRE (1988) Tactics, strategies and templets. Oikos 52: 3-18

Southwood TRE, May R, Hassel MP, Conway GR (1974) Ecological strategies and population parameters. Am Nat 108: 791-804

Specht MM (1985) Seasonality of aerial, litter and soil fauna in Dark Island heathland, South Australia. In: Greenslade P, Majer JD (eds) Soil and litter invertebrates of Australian Mediterranean-type ecosystems. WAIT School Biol Bull 12: 39-42

Specht MM, Specht RL (1992) Herbivory of leaves of Banskia oblongifolia, with or without overstorey cover. Proc VI Int Conf Mediterranean Climate Ecosystems (MEDECOS), Meleme (Crete) 1991, pp 22-27

Specht RL, Moll EJ (1983) Mediterranean-type heathlands and sclerophyllous shrublands of the world: an overview. In: Kruger FJ, Mitchell DT, Jarvis JUM (eds) Mediterranean-type ecosystems: the role of nutrients. Springer, Berlin Heidelberg New York, pp 41-65

Specht RL, Rayson P (1957) Dark Island heath (Ninety-Mile Plain, South Australia) III. The root system. Aust J Bot 5: 103-114

Stamatiadis S, Dindal DL (1990) The soil ecology of dry Mediterranean ecosystems with emphasis on the role of invertebrates in vegetal decomposition. Ann Mus Goulandris 8: 441-472

Stamou GP (1981) Population dynamics of Achipteria cf. italicus in the soil subsystem of an oakwood in Holomon Mt. PhD Thesis, University of Thessaloniki (in Greek with English summary)

Stamou GP (1986a) A phenological model applied to oribatid mites data. Rev Ecol Biol Sol 23: 453-460

Stamou GP (1986b) Respiration of Achipteria holomonensis (Acari: Cryptostigmata). Oikos 46: 176-184

Stamou GP (1995) Strategic responses of oribatid mites to the severity of the Mediterranean environment. In: Kropczynska D, Boczek J, Tomczyk A (eds) The Acari. Dabor, Warszawa, pp 295-304

Stamou GP, Asikidis MD (1989) The effect of density on the demographic parameters of two oribatid mites. Rev Ecol Biol Sol 26: 321-330

Stamou GP, Asikidis MD (1992) The effect of certain biotic factors on the demographic parameters of Scheloribates cf latipes (Acari: Oribatida). Pedobiologia 36: 351-358

Stamou GP, Iatrou GD (1993) Studies on the respiratory metabolism of Glomeris balcanica (Diplopoda, Glomeridae). J Insect Physiol 39: 529-535

Stamou GP, Pantis JD (1995) Responses of vegetation to the climate and anthropogenic severity of Greek Mediterranean ecosystems. In: Fantechi R, Peter D, Balabanis P, Rubio JL (eds) Desertification in a European context: physical and socio-economic aspects. Proc Eur Sch Climatol Nat Hazards. Directorate-General XII, Luxembourg, pp 321-338

Stamou GP, Sgardelis SP (1989) Seasonal distribution patterns of oribatid mites (Acari: Cryptostigmata) in a forest ecosystem. J Anim Ecol 58: 893-904

Stamou GP, Stamou GB (1996) Possible application of fuzzy system simulation models for bio-
monitoring soil pollution in urban areas. In: van Straalen NM, Krivolutsky A (eds) Bioindi-
cator systems for soil pollution. Kluwer, Dordrecht, pp 55-65

Stamou GP, Asikidis MD, Argyropoulou MD, Sgardelis SP (1993a) Ecological time versus standard clock time: the asymmetry of phenologies and the life history strategies of some soil
arthropods from Mediterranean ecosystems. Oikos 66: 27-35

Stamou GP, Iatrou GD, Sgardelis SP (1993b) Modelling the phenology of Glomeris balcanica. Ber
Naturwiss -Med Ver Innsb Suppl 10: 177-182

Stamou GP, Pantis JD, Sgardelis SP (1994) Comparative study of litter decomposition in two
Greek ecosystems: a temperate forest and an asphodel semi-desert. Eur J Soil Biol 30: 43-48

Stamou GP, Asikidis MD, Argyropoulou MD, Iatrou GD (1995) Respiratory responses of oribatid
mites to temperature changes. J Insect Physiol 41: 229-   -233

Stearns SC (1976) Life history tactics: a review of the ideas. Q Rev Biol 51: 3-47

Stefaniak O, Seniczak S (1981) The effect of fungal diet on the development of Oppia nitens
(Acari, Oribatei) and on the microflora of its alimentary tract. Pedobiologia 121: 202-210

Striganova BR (1972) Effect of temperature on the feeding activity of Sarmatoiulus kessleri
(Diplopoda) Oikos 23: 197-199

Terano T, Asai K, Sugeno M (1992) Fuzzy systems theory and applications. Academic Press, New
York

Testerink GJ (1983) Metabolic adaptations to seasonal changes in humidity and temperature in
litter-inhabiting Collembola. Oikos 40: 234-240

Trihas A, Legakis T (1991) Phenology and patterns of activity of ground Coleoptera in an insular Mediterranean ecosystem (Cyclades, Greece). Pedobiologia 35: 327-335

Tsiourlis GM (1990) Phytomass, productivité primaire et biogéochimie des écosystèmes méditerranéens phrygana et maquis (Ile de Naxos, Grèce). PhD Thesis, Université Libre de Bruxelles, Bruxelles

Turner BD (1983) Annual respiration and production estimates for collembolan and psocop-
teran epiphyte herbivores on larch trees in southern England. Ecol Entomol 8: 213-288

Vannier G (1978) La résistance à la desiccation chez les premiers arthropodes terrestres. Bull
Soc Ecophysiol 3: 13-42

Vannier G, Verdier B (1981) Critères écophysiologiques (transpiration, réspiration) permettant
de séparer une espèce souterraine d'une espèce de surface chez les insectes collemboles.
Rev Ecol Biol Sol 18: 531-549

Verhoef HA (1981) Water balance in Collembola and its relation to habitat selection: water content, haemolymph osmotic pressure and transpiration during an instar. J Insect Physiol 11:
755-760

Verhoef HA, Li KW (1983) Physiological adaptations to the effects of dry summer periods in
Collembola. In: Lebrun P, André HM, De Medts A, Gregoire-Wibo C, Wauthy G (eds) New
trends in soil biology. Proc VIIIth Int Colloq Soil Zool. Dieu-Brichart, Ottignies-Louvain-la-
Neuve, pp 345-356

Vitali-Di Castri V (1973) Biogeography of pseudoscorpions in the Mediterranean regions of the
world. In: Di Castri F, Mooney HA (eds) Mediterranean-type ecosystems. Origin and structure. Springer, Berlin Heidelberg New York, pp 195-306

Wallwork JA (1972) Distribution patterns and population dynamics of the micro-arthropods of
a desert soil in southern California. J Anim Ecol 41: 291-310

Wallwork JA (1982) Desert soil fauna. Praeger, New York, 296 pp.

Warburg MR (1987a) Isopods and their terrestrial environment. Adv Ecol Res 17: 187-242

Warburg MR (1987b) Haemolymph osmolality, ion concentration and the distribution of water
in body compartments of terrestrial isopods under different ambient conditions. Comp
Biochem Physiol 86: 433-437

Warburg MR (1992) Reproductive patterns in three isopod species from the Negev Desert.
J Arid Environ 22:73-86

Warburg MR (1994) Review of recent studies on reproduction in terrestrial isopods. Invertebr
Reprod Dev 26: 45-62

Warburg MR, Ben Horin A (1979) Thermal effect on the diel activity rhythm of scorpions from mesic and xeric habitats. J Arid Environ 2: 339-346

Warburg MR, Cohen N (1992) Population dynamics and longevity of Armadillo officinalis (Isopoda; Oniscidea), inhabiting the Mediterranean region of northern Israel. Pedobiologia 36: 262-273

Warburg MR, Rankevich D, Chasanmus K (1978) Isopod species diversity and community structure in mesic and xeric habitats of the Mediterranean region. J Arid Environ 1: 157-163

Warburg MR, Goldberg S, Ben-Horin A (1980) Scorpion species diversity and distribution within the Mediterranean and arid regions of northern Israel. J Arid Environ 3: 205-213

Warburg MR, Linsenmair KE, Bercovitz K (1984) The effect of climate on the distribution and abundance of isopods. Symp Zool Soc Lond 53: 339-367

Warburg MR, Cohen N, Weinstein D, Rosenberg M (1993) Life history of a semelparus isopod, Schizidium tiberianum Verhoeff, inhabiting the Mediterranean region of northern Israel. Isr J Zool 39: 79-93

Wernik AM, Penteado CHS, Cabral RA (1983) Respiratory metabolism of Leprodesmus dentellus and Sandalodesmus gasparae at different temperatures. Naturalia, (Sao Paulo) 8: 227-234

Westman WE (1981) Factors influencing the distribution of species of Californian scrub. Ecology 62: 439-455

Whittaker RH (1965) Dominance and diversity in land plant communities. Sciences 147: 250-260

Whittaker RH (1972) Evolution and measurement of species diversity. Taxon 21:213-251

Wood TG, Lawton JH (1973) Experimental studies on the respiratory rates of mites (Acari) from beech-woodland leaf litter. Oecologia 12: 169-191

Wooten RC, Crawford CS (1974) Respiratory metabolism of the millipede Orthoporus ornatus (Girard) (Diplopoda). Oecologia 17: 179-186

Wooten RC, Crawford CS (1975) Food, ingestion rates and assimilation in the desert millipede (Girard, Diplopoda). Oecologia 20: 231-236

Wooten RC, Crawford WA, Riddle WA (1975) Behavioural thermoregulation of Orthoporus ornatus (Diplopoda: Spirostreptidae) in three desert habitats. Zool J Linn Soc 57:59-74

Yeilding L (1977) Decomposition in chaparral. In: Mooney HA, Conrad LE (eds) Proc Symp on the environmental consequences of fire and fuel management in Mediterranean ecosystems, Palo Alto, California, pp 419-425

Young SR (1979) Respiratory metabolism of Alaskozetes antarcticus. J Insect Physiol 25: 361-369

Zachariassen KE, Andersen J, Maloiy MO, Kamau MZ (1987) Transpiratory water loss and metabolism of beetles from arid areas in East Africa. Comp Biochem Physiol 86: 403-408

Species Index

A. officinalis   6, 73
A. oudemansi   37, 73
A. t. judaicus   22, 23, 62, 65, 67, 68, 73, 75,
    84
A. vulgare
Achipteria oudemansi   30, 72
Acinopus subquadratus,
Arbutus unedo   8, 98
Archispirostreptus tumuliporus judaicus
    22, 43, 44, 60, 97
Armadillidium vulgare   59, 97
Armadillo officinalis   64
Asparagus aqutifolius   8
Australiosoma castaneum   112

Badisis ambulans   46
Balota acetabulosa   13
Banskia oblongifolia   45
Buthotus judaicus   42, 97

Carterus calydonious   92
Carterus calydonius   78
Cephalatus follicularis   45
Ceratophysella engadinensis   107
Cistus incanus   8
Cryptopygus antarcticus   32

Dailognatha quadriocollis   78

Erica arborea   8
Folsomia quadrioculata   110
Folsomides angularis   26

G. balcanica   34, 35, 36, 37, 39, 46, 47, 49,
    52, 53, 56, 62, 68, 69, 70, 73, 84, 97, 111
G. leucoloma   59, 67
G. marginata   52
Glomeris balcanica   30, 60, 97
Glomeris marginata   47, 60
Graphognathus leucoloma   59

Isotomurus palustris   107
L. cynarae   99
Larinus cynarae   99

M. tublaghia   45
Meneris tulbaghia   45
Metaponorthus sp   77
Metoponorthus pruinosus   73, 92

Nebo hierochonticus   42

O. cincta   22
O. meridiatus   31, 36, 37, 110
O. minimus   103
O. moreletii   47
Olodiscus minimus   103
Ommatoiulus moreletii   23, 42, 66, 68, 75,
    84
Onopordum illyricum illyricum   99
Onychiurus meridiatus   30, 51, 107
Onychiurus zschokkei   96
Orchesella cincta   19
Orchesella irregularilineata   107
Orthoporus ornatus   23

P. allifera   72, 73
Parisotoma octooculata   32
Phillyrea media   98
Philoscia muscorum   73, 92
Philyrea media   8
Picris echioides   110
Pilogalumna allifera   30, 72
Pinus halepensis   98
Pistacea lentiscus   8
Porcellio olivieri   97
Pseudosinella albida   107

Q. coccifera   8, 12, 13
Quercus coccifera   6, 73, 98
Quercus ilex   96

S. cf. latipes   72, 73, 112, 113
S. latipes   103
S. tiberianum   47, 65, 67
Scheloribates   110
Scheloribates cf latipes   30, 65, 71
Scheloribates latipes   51, 101, 111
Schizidium tiberianum   42, 64

Scorpio maurus fuscus   97
Subisotoma variabilis   26

Thymus capitatus   13
Tomocerus minor   20

Xanthoria parietina   110
Xenylla sp.   107

Subject Index

Abrasion   21, 23
Acclimation   31, 35-40
Activity
    daily   41-42
    locomotory   19, 21, 50-51
    patterns   41-48, 84, 94
Adaptation
    ancestral   22, 59, 119
    apotypic   62, 119
Aestivation   69-70, 120-121 123
Age at maturity   60, 62, 69
Age distribution   63, 71
Aggregation   7, 13, 43, 113
Anhydrobiosis   25-27
Asexual reproduction *see* Parthenogenesis
Assimilation coefficient   52, 53

Batha *see* Phrygana
Breeding
    period   42, 47-48
    success   66-68, 73-74
Brood protection   67-68, 120

Caloric content   52
Chaparral *see* Maquis
Climate   1-2
Clutch size   60, 67-68, 74, 124
Coastal sages *see* Phrygana
Coevolution   45, 63
Colonisation   103, 106, 122
Community structure   7-13, 91-103
Compensation
    metabolic   34-36, 39, 95, 123
Conformism   120-124
Conservatism   120-124
Constraints
    evolutionary   59, 121, 124
    physiological   59, 110, 119, 121
Convergence
    ecological   9
    hypothesis   11

Cost
    of maintenance   34, 36, 39, 52-53, 120, 124
    ontogenetic   54
Cryptobiosis   25

Degradation   4, 9-13, 117
Demography
    characteristics   59-71, 88-90, 120-124
    parameters   50, 54-57, 66, 87, 111-115
    vector   87
Density
    dependence   14, 113-115, 118-119, 121-124
    dependent foraging   112-115
    dependent movement   115
Diapause   69, 119, 123
Disasters   86
Distribution
    of abundance   7-8, 107-110
    geographical   7-11, 91
    horizontal   97
    of rainfall   1
    of reproductive effort   65-67
    of resources   43, 63, 117
    vertical   96-97
Dormancy   60, 69, 72-73
Drought
    avoidance   94, 97, 122
    hypothesis   12
    index of   1
    resistance to   18, 72, 73, 94
    response to   18, 24-27, 85, 96-97

Ecdysis   19-24, 110, 113
Ecological time   79-84
Ecomorphosis   19, 25-27
Effort
    reproductive   61, 65-68, 82-83, 103, 120, 122
Egg
    chamber   67
    investment in   60, 68, 71

production   37, 46-48, 56-57, 66-75, 121-
    124
Energy
    budgets   52
    transformation   31, 32, 39, 120
Exploitation
    habitat   98
    of land   10, 14, 53, 86
    of resources   8, 53, 113-115, 124

Genetic structure   99

Fasting   18-19, 26-27
Fecundity   61, 87, 111, 114
Feeding
    activity   44-46, 48-51, 113, 124
    habits   110-112, 124
Fire   100-106
Fitness   118
Food
    selection   110, 122
    web   45-46, 52
Fuzzy modeling   87
Fynbos *see* Maquis

Generation time   60-63, 120-124
Grazing   12, 106-110

Habitat
    selection   22, 24, 40, 100, 103, 106-115, 122
    templets   118-119
Heat budgets   37, 60, 62, 71, 87

Impact
    anthropogenic   9-13, 99-110
Iteroparity   63-67, 72, 123

Landscape   10-13, 91, 96-99, 117
Life cycle
    development   71-75, 118-123
    synchronisation   37, 53, 69, 73-77, 84,
        122-124
Life history
    characteristics   59-71, 82
    models   14, 118-119
    strategies   29, 39, 53, 59, 118-124
Life span   60-62
Litter
    cconsumption   52
    decomposition   15, 94-95
    production   15
Longevity   55, 124

Mallee *see* maquis
Management
    history   10

practices   10, 117, 123
Maquis   2, 5, 7-9, 12-13
Matoral *see* Maquis
Metabolism
    constraints   119-124
    respiratory   29-40
Metapopulation   98-99, 115, 122, 124
Migration
    horizontal   41, 56, 98, 99
    orthokinetic   18
    vertical   27, 56, 96, 97, 103
Mortality   20, 33, 53-54, 65, 82, 87, 120, 124

Nutrients   14-15, 96

Osmolalitiy   19-24, 122
Osmoregulation   20-24
Oviposition
    Circumscribed   65, 69, 98, 121, 124
    patterns   46-48
    period   44, 56, 61, 72
    risk   71

Parental care *see* Brood protection
Parthenogenesis   63-64
Partitioning of resources   110, 115
Past
    thermal   55-56
Permeability
    cuticular   21-24
Phenogram   77-81, 84-86, 121
Phoresy   98, 103, 119, 123
Phrygana   9-10
Plateau
    density   75
    mortality   54
    thermal   39, 51, 120
Pollination   44-45
Population
    dynamics   71-73, 77-90, 121
    reserves   37, 71, 120
Potential
    reproductive   55, 88, 106, 118
Precocity   65, 67, 72, 82, 120, 122
Predation   42, 67, 99, 112-113
Predator avoidance   42, 119

Q_{10} 34, 39, 51, 120
Quiescence   53, 69, 98, 119-123

Recovery
    of activity   20
    metabolic   27
    of population   21, 82, 86-90, 101-106,
        121-122
    time   64

Renosterbos *see* Phrygana
Resilience   10-15
Respiration
    biomass   31
    coefficient   52
    -mass relationship   30-32
    -temperature relationship   32-35
Restoration   64, 106
Rhythms   41, 69

Selective forces   12, 107, 112, 119, 123
Semelparity   62-65
Severity
    response to   117, 120
    environmental   11, 117
Sex ratio   64
Soil profile   5-6
Species
    diversity   7-10, 92
    core   8, 107-110
    satellite   8, 107-110
Strategy
    adaptive   29, 103, 117-124
    A-strategy   123
    life history   53
    maxithermal   41
    R-C-S   14
    reproductive   62-67
    respiratory   39
    r-K strategy   118-121
    strategic thinking   29, 118-119
    tolerence   19

Succession   107
Survival curve   50, 55, 82, 112
Synchronisation tactics   69-71
Synchrony
    generational   121, 124

Thelytoky *see* Parthenogenesis
Thermoregulation
    behavioural   41, 122
    evaporative   24
    metabolic   55-56, 65
Threshold
    density   114
    environmental   44, 69, 122
    moisture/humididty   17, 49, 124
    temperature1   33, 49, 50, 53, 70, 119, 124
Tolerance   19, 25-27, 84
    range of   34, 120, 126
Tomillares *see* Phrygana
Transition matrix   87
Translocation
    biomass   14
    nutrient   14
    water   24
Transpiration   21-26, 122

Voltinism   69, 121

Water
    balance   18-22, 122
    deficit   4
    relations   17-27, 122, 124

Springer
and the
environment